The World's Greatest
TRACTORS

The World's Greatest
TRACTORS

John Carroll

Grange
BOOKS

Published in 1998 by
Grange Books
An imprint of Grange Books Plc.
The Grange
Kingsnorth Industrial Estate
Hoo, nr. Rochester
Kent
ME3 9ND

ISBN 1 84013 153 5

Printed in Italy

PAGE 2: The Massey-Ferguson 390 tractor is from the
300 series range and is turbo-diesel-powered.

PAGE 3: The John Deere 6800 tractor is one of the
models that comprises the 6000 range.

RIGHT: Fiat tractors, such as this 315 of 1966, were
frequently sold abroad but under other names. For
example, Oliver of the United States sold Fiats as the
smaller tractors in its range.

OPPOSITE: A Case 4230.

Contents

Acknowledgments

The author and photographers are indebted to the owners and manufacturers who allowed their tractors to be photographed. Thanks also go to the organizers of the following agricultural and vintage shows; HMT Historisch Festival, Netherlands, Great Dorset Steam Fair, England, La Locomotion en Fête, France, and Western Minnesota Steam Threshers Reunion, U.S.A.

Photographs by Andrew Morland, Ian Clegg, John Carroll and Garry Stuart.

ABOVE
A Massey-Ferguson 399 at work in the rolling Malvern Hills of Worcestershire, England.

RIGHT
A John Deere 7800 tractor complete with a John Deere-manufactured plough.

Introduction

A 1910 Ford experimental motor cultivator incorporated Model 'T' parts.

It is an acknowledged fact that the era of mechanically propelled transport began in 1769 when Nicholas Cugnot built a three-wheeled vehicle powered by steam propulsion – the first machine specifically designed for haulage – capable of 4mph (6km/h) that ran in the streets of Paris and carried four passengers. A Welsh inventor, Oliver Evans, who had emigrated to America and lived in Maryland, produced an elementary steam wagon in 1772. He explored the possibilities of applying steam power to propel a vehicle and in 1787 was granted permission to proceed. His wagon never made it as far as production but he did build a steam-powered dredging machine that he rigged to drive from its place of manufacture to the River Schuylkill and thence to Delaware. In 1788, a vehicle built along similar lines – the Fourness – was built in Britain and by 1831 the idea was proven and men such as Sir Charles Dance and Walter Hancock operated a number of steam coaches on regular routes, the latter's machines being capable of up to 20mph (32km/h). The railway age boomed and the mechanization of road transport and other machinery, including agricultural, was a natural progression and

experimentation towards practical methods of achieving this continued in both Europe and the United States.

In the closing years of the 19th century, vehicles powered by internal combustion engines began to make an appearance and names like Benz, Daimler, De Dion, Panhard and Peugeot became prominent in Europe, while in Britain Albion, Dennis, Humber, Napier, Sunbeam and Wolseley came to prominence. The first British commercial vehicle that was both viable and practical was made by Thornycroft in 1896, the same year that the 'Red Flag Act' was repealed by Parliament. This act required operators of vehicles powered by internal combustion engines to employ a man to walk in front of them with a flag, warning of the danger, and the repeal of this odd legislation paved the way for the development of both the steam and internal combustion engines. A year earlier, Richard F. Stewart of Pocantico Hills, New York

An archive picture of a Model 'G' 1916 Sanderson Universal with a 20–25hp engine, hard at work pulling felled trees.

produced his first truck with a 2-hp Daimler engine and internal gear-drive and two years later began to produce trucks for sale which were steam-powered with marine-type steam boilers and vertical engines. Within a few years Thorneycroft had produced the world's first articulated truck. The British Army was quick to realize the potential of such machines and acquired some by 1899 for use in the Boer War in South Africa. It used steam traction engines for hauling artillery. Leyland vehicles appeared in 1896

under the auspices of the Lancashire Steam Motor Company whose first vehicle was a van with oil-fired boiler and a two-cylinder compound engine. The manufacturers of steam traction engines began to build steam lorries and one of them, Foden, was to become the world's largest steam vehicle producers.

In 1897 the Daimler Motor Company of Coventry, England offered a petrol-engined commercial vehicle. It was designed by Panhard and powered by a Daimler internal combustion engine.

From this vehicle it progressed to building other similar machines while across the Atlantic the first Mack truck was rolling out of a Brooklyn works. The company had been established by five brothers of German parentage who had formerly operated a Brooklyn smithy. The smithy was gradually turned over to the production of trucks and its first is reputed to have travelled a million miles. In 1904 the brothers built a charabanc which they named the Manhattan and by 1905 had become sufficiently successful to make the transfer to Allentown, Pennsylvania where they introduced the Model AC. This was the truck that earned the Bulldog Mack nickname and it was a four-cylinder petrol-engined truck based on a pressed steel chassis frame. Transmission was by means of a three-speed gearbox through a jackshaft which had chain-drive to the rear wheels which were fitted with solid tyres. The truck featured a cab which was an unusual feature at the time and was of a bonneted design. The AC was supplied in significant numbers to the British Army in France and this is where the 'Bulldog' tag began to stick.

Another Mack landmark was the predecessor of the 'cab-over-engine' design when it was realized that if the driver sat over the engine it was possible to incorporate a longer load-bed in a chassis of the same length. The fledgeling Mack company also developed the Junior model, a two-ton truck intended for lighter duties which in many ways was the forerunner of the delivery van, while other companies developed more specialized machines for farm use. From common beginnings, trucks, buses, cars, vans and, of course, tractors evolved and subsequently diverged and, within the span of a century, tractors have progressed from being primitive and unreliable to utterly sophisticated and precise machines. The latest innovations include turbo-diesel engines, four-wheel-drive transmissions and operator comforts such as vibration-proof, air-conditioned cabs.

Four-wheel-drive systems, and the engineering behind them, have come a long way since they were first conceived. Four-wheel-drive differ from two-wheel-drive vehicles simply because they use a different transmission system to drive two axles rather than just one. Traditional four-wheel-drive vehicles tend to use a pair of live axles that are attached to the chassis by either coil or leaf springs and to drive the differential in each axles requires a gearbox with two drive-shafts, one facing the front axle, the other

The first International Harvester wheeled, diesel-powered tractor appeared in 1934 and was designated the WD-40. It was powered by a four-cylinder engine and the numerical suffix indicated that its power was in the region of 40bhp – in fact, 44bhp.

LEFT
This David Brown VAK 1A tractor is undergoing manufacturer's tests.

OPPOSITE
The Lanz Bulldog was a popular tractor from its inception and soon there were a range of Bulldogs including crawler variants. The Bulldogs survived the war and in post-war Germany their production was resumed, hydraulic lifts being fitted in the fifties.

the rear. The part of the transmission that achieves this is known as the transfer box and to allow for better gear ratios for off-road use, the transfer box in most 4x4s has two speeds – high ratio and low ratio – and it is the latter that is used off-road. In many 4x4s these are selected by means of an additional gear lever and the term 4x4, incidentally, is of military origin being the way vehicles are classified, the first 4 referring to the number of wheels a particular machine has and the second 4 to how many are driven; so it is equally possible to have 6x6 and 6x4 vehicles, as in some trucks and machines.

There are two main types of four-wheel-drive systems, full-time and part-time. Full-time means that the vehicle is constantly driving on all four wheels while part-time systems allow four-wheel-drive to be selected for difficult conditions and it was this

system that was widely proven first. The reason why part-time systems need drive to the second axle selecting is because of what is referred to as 'wind-up'. Although there are differentials in each axle there is no differential in the gearbox, so both shafts turn equal numbers of revolutions. While operating on loose and muddy surfaces the wind-up is scrubbed off, on surfaces where traction is better it is not, so the transmission tightens up as wheels rotate at different speeds when cornering. As the transmission tightens up it will increase tyre wear, make changing gear more difficult and eventually damage the gearbox. For a permanent four-wheel-drive system to work it was necessary to devise a gearbox that would not suffer from transmission wind-up. The way this was resolved was to fit a differential device in the gearbox which would allow both drive-shafts to turn at different speeds. This, however, creates its own problems because a differential means that the drive to it will go to wherever the wheels turn most easily. On a full-time four-wheel-drive system this means that it is possible to have three wheels on firm ground and one in mud spinning so that the vehicle is stuck! (The comparable situation on a selectable four-wheel-drive system is that one wheel on each axle must be spinning to be stuck.) To overcome this, the 'diff lock' – as the differential lock is often referred to – was developed. By means of a lever, the driver of a permanent four-wheel-drive tractor can engage the diff lock which means that the centre or gearbox differential is in effect overridden, putting the machine back on a par with the part-time system. Some systems have, in recent years, benefited from a viscous coupling that engages the diff lock automatically when a wheel begins to spin. To further enhance a 4x4's off-road ability, it is possible to fit locking differentials to either or both front and rear axles. Such differentials are manufactured by a number of companies around the world.

Although live axles – a tube that contains the differential and half-shafts – are found on many four-wheel-drive machines, it is not uncommon to see independently-sprung systems. These feature a differential mounted on the chassis with unenclosed shafts running to the hubs: Steyr Puch used this system for the Haflinger and Pinzgauer vehicles. The other important factor about axle type and suspension is that, as well as guaranteeing a degree of comfort for the driver, it ensures that the wheels stay in contact with the ground – vital if traction is not to be lost. The movement of the axles over rough ground is described as axle articulation and the greater articulation an axle has, the rougher the ground the vehicle can cross.

Driving a tractor both on- and off-road calls for different techniques than for other machines. To give a tractor sufficient ground clearance, designers have to build them to stand tall: the wheels tend to be of 15 or 16 inches (406mm) in diameter and the engine and gearbox components are mounted above the line of the axles. This means that the centre of gravity of a tractor is high and it can be turned over if sufficient care is not exercised. Descending steep hills must be done using the engine and transmission to brake, rather than the brake pedal, because it is easy to lock the wheels, causing the machine to slide out of control. Inclines should be ascended and descended perpendicular to the slope to prevent them from rolling over. In deep water, care must be taken to avoid damaging the machine through ingress of water into the engine, transmission and axles. Other aspects of off-highway use can be hazardous too, such as the use of winches in the timber industry. Winching requires careful observation of a number of basic safety rules: a winch operator should wear gloves to protect his or her hands from broken wire strands and should always stand clear of a taut cable in case it snaps. The forces are such that if it were to snap it could quite likely injure or kill an operator. For this reason, no one should step over a taut cable. Other dangerous operations can involve power take-off (PTO) shafts and drives and, in the past, the pulleys for belt-drives to implements and machines.

Tractors are applied to every area of farming and have become increasingly sophisticated in recent years. At the same time there are inherent dangers in their operation and this is illustrated by a report of seven people being killed in tractor accidents in Yorkshire, England between 1988 and 1993, while a further 21 were seriously injured. An obvious example of such an accident is where a tractor is driven across the gradient of a hill and rolls over and tractors are often used to drive other machinery which have their own hazards. It was also noted that a significant number of those involved in accidents were in the 18–25 age group, suggesting that inexperience may be a contributory factor. Seeking to reduce these figures, the Health and Safety Executive, a British official body investigating over 1,000 tractor accidents, launched a campaign called Tractor Action. It incorporated a video and training pack and was aimed at colleges, lecturers and others involved in work-based training. If these figures are extrapolated to give a national and, indeed, international overview, it becomes instantly apparent that driving and manipulating tractors is not to be undertaken lightly.

OPPOSITE
White bought Oliver in 1960 and continued to produce the successful and popular Oliver 55 models throughout the next decade.
BELOW
A French Someca tractor at work at a vintage agricultural show.

Chapter 1
PATENTS PENDING 1900–1920

The mechanization of farming can be traced back to such events as Jethro Tull's invention of the seed drill and Andrew Meikle's development of a machine that later became the mechanical thresher. The need for a mechanical source of power to drive early machines was clearly understood and some experimentation occurred. Certain experiments with steam power proved worthwhile and soon machines such as Thomas Newcomen's steam pump were operational. The concept was refined by James Watt and Nicholas Cugnot, the latter having built a steam artillery tractor in 1769. Cugnot's tractor was the first machine specifically designed for haulage but was not immediately popular. The advent of the railway locomotive again focused attention on the possibilities of steam-powered machinery, independent of rails. Gradually the steam traction engine became more refined and a practical proposition for use in farming. However, in the main, the steam traction engine was reserved for providing mobile stationary power to drive threshing machines and similar. The obvious exception were the pairs of ploughing engines employed in field cultivation. When ploughing fields, these

OPPOSITE
Over the years, many of the smaller tractor manufacturers have been absorbed by others. Advance Rumely acquired Aultman and Taylor and were themselves acquired by Allis-Chalmers in 1931.

LEFT
An Allis-Chalmers 20–35hp tractor. The steel wheel rims are plainly evident but the next major advance in tractor technology would be to use pneumatic tyres. Allis-Chalmers went so far as to hire racing drivers to demonstrate their capabilities.

relied on drawing a plough backwards and forwards between them using a winch system rather than simply by towing a plough behind and in many areas this method endured until the 1930s. The great size of farms in North America and South Africa meant that the need for further mechanization was all the more necessary. Progress came in the shape of the internal combustion engine and associated developments of the time and from the workable internal combustion engine it was but a short step to a practical agricultural tractor.

John Charter built petrol engines in Stirling, Illinois and manufactured a tractor by fitting one of his engines to the chassis and wheels of a steam traction engine. The machine was put to work on a wheat farm in South Dakota in 1889, it was a success, and Charter is known to have built several more machines to a similar specification. By 1892, a number of other fledgeling manufacturers were beginning to produce tractors powered by internal combustion engines. John Frohlich built a machine powered by a Van Duzen single-cylinder engine and with his own design of transmission and system mounted the engine on the rolling chassis of a steam engine. Frohlich had experience in agriculture, having worked as a threshing contractor, so was fully conversant with the requirements of mechanized harvesting. He bought a wagon in which to live and a large Case thresher and moved the whole ensemble by rail to Langford, South Dakota. It is reported that hundreds turned out to see him start up the machine where over a period of seven weeks he carried out wheat threshing continuously. His machinery suffered no breakdowns and convinced many of the benefits of mechanization. Frohlich gained backing from a group of Iowan businessmen to form the Waterloo Gasoline Traction Engine Company which later dropped the word traction from its name and manufactured stationary engines until it produced another tractor in 1912. It was a similar story for numerous other companies: Case experimented with gasoline engines based on a design by William Paterson before returning to steam. Other tiny companies built prototypes but achieved little more until the Huber Company of Marion, Ohio purchased the Van Duzen Engine Company and produced a batch of 30 tractors. A couple of other firms also started up around this time, which included the Otto Gas Engine Company and Kinnard-Haines of Minneapolis.

OPPOSITE
A Hart-Parr 28-50. Hart-Parr, one of the pioneer manufacturers, redesigned its machinery for the post-war boom. The smaller 12-25 was one of the new models and featured a horizontal twin-cylinder engine.

LEFT
The Samson Iron Horse was a tractor briefly manufactured by General Motors at Janesville, Wisconsin.

ABOVE
The Heider C 12-20 tractor made
by Rock Island Plow of Illinois was
the first tractor with more than two
forward gears to be tested in the
Nebraska Tractor Tests.

OPPOSITE
A Twin City 1910 40hp four-
cylinder. Twin City Tractors,
Minnesota Steel and Machinery, the
Minneapolis Threshing Machine
Company and the Moline
Implement Company all merged to
form Minneapolis-Moline.

Two other companies, Deering and McCormick, built self-propelled mowers at this time, all contributing to the speed with which mechanized farming in America was increasing.

Similar tentative steps were being taken in various European countries, although economic conditions were different there, labour being more plentiful and cheaper than in the United States. In Germany, Adolf Altona built a tractor powered by a single-cylinder engine that used chain-drive to the wheels. This machine was not wholly successful, but progress was later to be assured as a result of Doctor Rudolph Diesel's experiments with engines. Diesel's engines used high compression to ignite and were of course the beginnings of what is now known as the diesel engine. In England, Hornsby of Lincoln was producing tractors in the 1890s from a licence-built version of someone else's machine and Petter's of Yeovil, and Albone and Saunderson of Bedford would later also build tractor-type machines.

Dan Albone was a bicycle manufacturer with little, if any, experience of the steam propulsion industry and consequently approached the task of producing a tractor from a different viewpoint. Borrowing some ideas from the infant automobile industry, he built a tractor which he named after a river near his home, the Ivel. His machine had only three wheels but was practical and suited to a variety of farm tasks. It was a success and production commenced. Some were exported and no doubt the company would have developed into a major force in the industry except for Albone's untimely death in 1906. The company seemed to falter without his involvement and ceased production in 1916. Other manufacturers were coming on the scene by this time and included Ransome's of Ipswich while Petter produced its Patent Agricultural Tractor in 1903. Marshall and Daimler built machines and sought to export them, a Marshall machine being exhibited in Winnipeg, Canada in 1908. In Germany, Deutz introduced a tractor and motor plough of advanced design in 1907 while Deutsche Kraftplug, Hanomag, Pohl, and Lanz were also involved in tractor and powered agricultural machine manufacture. In France, De Souza and Gougis were just two to be represented at a tractor trial at the National Agricultural College at Grignon, near Paris, at which event tractors undertook a variety of voluntary and compulsory tests. Elsewhere in Europe there was progress too: Munktell in Sweden made a tractor in 1913, Pavesi in Italy produced the Tipo B and there were three designs from a Russian company before the onset of the First World War. On the other side of the world, the McDonald tractor was unveiled in 1908 in Australia. This was followed by machines from Jelbart and Caldwell-Vale.

The economic conditions prevailing on both sides of the Atlantic prior to the First Word War meant that the bulk of tractor production was focused in the United States. Designs began to appear specifically intended for American prairie cultivation which included tractors for drawbar towing of implements, especially ploughs. The International Harvester Corporation was formed in 1902 through the merger of McCormick and Deering, and Avery, Russell, Buffalo-Pitts and Case all built experimental machines in the first decade of the 20th century. Case was operating in 1911 and by 1913 offered a relatively practical tractor powered by a gasoline engine. Another early tractor was manufactured by two engineers, Charles Hart and Charles Parr, but like some of the preceding models their first attempt was a heavy and ungainly monster. However, they went

on to produce more practical machines and these included the 12-27 Oil King. The practical value of such machines was apparent from the level of exports being made around the globe to countries as distant as Russia, South America and Australia.

The youthful International Harvester company marketed two ranges of tractors, the Mogul and Titan models, the former being sold by McCormick and the latter by Deering dealers. Rumely of La Porte, Indiana, who was primarily a manufacturer of threshers, drove its first tractor out in 1909. One of its employees had perfected a carburettor for kerosene or paraffin and a model known as the Oil Pull was gradually improved and remained Rumely's main product until the company was sold to Allis-Chalmers in 1931.

Experimentation continued right up to the outbreak of the First World War. The Heer Engine Company of Portsmouth, Ohio produced a four-wheel-drive tractor in 1912 and the Wallis Tractor Company a frameless model – the Cub – in 1913. The popularity of tractor trials grew, their purpose being to evaluate performance and make realistic comparisons between the various models. Following the success of the Canadian Winnipeg Trials of 1908, these continued as a regular event until 1912. A small tractor trial was held in Great Britain in 1910 while in the United States, similar trials were held in Nebraska.

The outbreak of the First World War in Europe was to have far-reaching effects, both in terms of the economics of farming and the production of tractors. Up until then, Britain was in the luxurious position of obtaining much of its food from its Empire, which stretched across the world. As a result, Britain had less need for self-reliance and farming was in a somewhat depressed condition. However, the war suddenly demanded a massive number of horses and men when it was Germany's aim to starve Britain into surrender. The coalition government of the time therefore instituted policies to encourage domestic food production which ranged from ploughing and cultivating land that had been allowed to stand fallow to increasing mechanization of farming in an attempt to produce more with fewer people. Some of the pre-war tractor producers began to turn their attention to the war effort while Ruston Hornsby of Lincoln, for example, became involved in tank experiments.

Saunderson tractors were still in production and Weeks-Dungey entered the market in 1915 and there was an obvious need to import tractors from America. Soon after, the Austin Motor Company offered the Peoria from Illinois which was known in England as the Model 1 Culti-Tractor. The Big Bull was marketed as the Whiting-Bull and a Parrett model was renamed the Clydesdale in an obvious reference to the horse-drawn plough. Another famous tractor of the time was the Waterloo Boy sold in Britain as the Overtime by the Overtime Farm Tractor Company based in London. The International Harvester Company marketed models from its range that it perceived to be most suited to British conditions. These were the Titan 10-20 and the Mogul 8-16. Fordson tractors also became established at this time and are discussed later on (see p. 82).

Government policy paid off and by 1918 the wheat harvest exceeded that of 1916 by approximately 50 per cent and the production of other crops, including barley, oats and potatoes, had also increased significantly. While Britain at least escaped widespread war-damage to the landscape this was not the case in Europe, where the Western Front was completely destroyed. The same could be said of the British workforce which had been decimated in the service of king and country and this made the tractor and mechanized farming all the more necessary.

Some other European nations involved in the war also began to see the value of tractors as a means of increasing farm productivity. The French imported American tractors and gave them more French-sounding names while elsewhere Globes were renamed Allis-Chalmers 10-18s, Czars became Bean Tracpull 6-10s and Le Gaulois the Galloway Farmobile 12-20. In Italy its motor industry undertook experiments with tractors of its own design and by 1918 had produced a successful Fiat tractor, the 702, and the innovative Pavesi 4x4 with articulated steering.

The Germans were blockaded by the allied navies by 1917 and were forced to rely on increasing their own production of food. They seem to have been more reliant on horse ploughs than other European nations though the production of motor ploughs by Lanz, Stoewer and Hansa-Lloyd continued. Lanz manufactured the Landbaumotor, Stoewer the 3S17 and 6S17 models and Hansa-Lloyd the HL18.

The United States at first remained neutral in the First World

war with Mexico that had begun in 1846 and led to the American occupation of Veracruz. America declared war on Germany and its allies on 6 April 1917 in spite of the fact that its military was not as well prepared as might have been expected. The Democrat president, Thomas Woodrow Wilson, made a plea to industry requesting a concerted effort and, as a result, production reached an all-time high. Across the Atlantic, the tractor market was growing and the realization that smaller tractors were a practical proposition changed the emphasis of the industry and threatened some of the older established companies. The Minneapolis Steel and Machinery Company produced the Bull tractor for the Bull Tractor Company which was a small machine based around a triangular steel frame. It had only one driven wheel to eliminate the need for a differential, while a single wheel at the front steered the machine and a third simply freewheeled. An opposed twin engine produced up to 12hp and the transmission was as basic as the remainder of the machine, having a single forward and single reverse gear. Initially the Bull tractor sold well but, as the limitations and faults within the machine's design and construction became apparent, sales declined and little was heard of the company after 1915. The Fordson was destined to be an altogether more practical version of the same idea. Case was experimenting with smaller machines and produced the four-wheeled 9-18 model in 1916. It was this machine that in many ways established Case as a major tractor manufacturer. Other smaller companies, which included Ebert-Duryea, Lang, Fagiol, Kardell, Michigan, Utility and Happy Farmer experimented and innovated but were never realistic long-term propositions.

The First World War ended in 1918 and despite the major socio-economic upheavals that came in its wake there had been huge progress in the engineering and automotive industries. Many of the prototypes of pre-war days had been transformed into practical and functional vehicles. Tyre technology, transmissions, engines, and more, had all progressed and tractor manufacturers were able to further exploit these developments.

War, many seeing it of primary concern only to Europeans. But after the holder of the Blue Ribband, the liner Lusitania, was sunk by a German U-20 off the Old Head of Kinsale, Ireland, with the loss of 1,198 lives in 1915 and the torpedoing of non-combatant shipping in the Atlantic, greater anti-German feeling was aroused and ultimately precipitated the United States into the war on the side of the Allies.

For manufacturers of every type of motor vehicle there was the increasing prospect of sales to the military. It was no secret that there were experiments with automobiles, as a result of the

OPPOSITE
Henry Ford built his first tractor in 1915. He aimed to produce an affordable tractor which would do for farmers what the Model T had done for motoring in general. Together with members of his staff, Ford had designed what would later become the Fordson F model of which this is a 1917 example. Its secret was that it was of a stressed cast-iron frame construction. This frame contained all the moving parts in dust-proof and oil-tight units which eliminated many of the early tractors' weaknesses.

LEFT
The International Harvester Company exported tractors to Britain where it marketed models from its range deemed most suited to British conditions. These were the Titan 10-20 and the Mogul 8-16. This is a 1919 International Harvester Titan 10-20.

Chapter 2
MASS-PRODUCTION 1920–1940

Prosperity followed war and tractor manufacturers around the world mushroomed. Many were tiny companies with little chance of success but the twenties was a time when mass-produced machines were gaining sales everywhere. Henry Ford's Fordson tractor sold in vast numbers, achieving some 75 per cent of total tractor sales. It was cheap to produce and affordably priced, making it accessible to a greater number of farmers. While many of the tiny manufacturers struggled on, with production in handfuls, the tractor market was developing into a struggle between Fordson, International Harvester, Case and John Deere.

As the initial wave of prosperity in the early twenties abated, Fordson prices were cut to maintain sales and IHC replied by offering a plough, at no extra cost, with every one of its tractors sold. This had the desired effect of shifting all IHC stock, allowing it to introduce the 15-30 and 10-20 models in 1921 and 1923 respectively. These machines were to provide Ford with a level of competition not previously experienced. The new IHC models were of similar construction to the Fordson, built around a

stressed cast frame but incorporating a few details that gave it the edge on the Fordson. These included magneto ignition, a redesigned clutch and a built-in power take-off, setting a new standard for tractors. In 1924 it went one better with the introduction of the Farmall, the first genuine row crop tractor which could be used for ploughing as well as turning its capabilities to cultivation of rows of cotton, cereals and other crops.

At this time Case also introduced a cast frame tractor and, although the engine ran across the frame, the model proved popular and endured until the thirties. John Deere offered its own interpretation of the cast frame tractor with the Model D of 1924. It was powered by a two-cylinder kerosene/paraffin engine and had a two forward and one reverse speed gearbox. From this basic machine a production run of sequentially upgraded models continued until 1953. John Deere introduced a row crop tractor in 1928 known as the GP while Hart-Parr, one of the pioneer manufacturers, redesigned its machinery for the post-war boom. The 12-25 was one of the new models and featured a horizontal

RIGHT
A 1931 International 10-20. Three models carried International Harvester through the twenties: the 10-20, 15-30 and Farmall Regular. These were refined and redesigned for the thirties.

OPPOSITE
An Allis-Chalmers WC tractor made in Wisconsin, U.S.A. It was introduced in 1934, the first tractor designed for pneumatic tyres, although steel rims were available as an option.

twin-cylinder engine. Slightly later, these tractors were offered with an engine-driven power take-off but the full import of this was not realized until later.

The downward economic trend that followed the boom led to numerous closures and mergers within the tractor industry. Case bought Emerson Brantingham while Advance Rumely acquired Aultman Taylor while continuing to refine its Oil Pull line of tractors. It was itself taken over by Allis-Chalmers in 1931 while Avery went into liquidation and Kinnaird-Haines ceased production.

The Nebraska Tractor Tests were established as a way of determining a tractor's capabilities and preventing its manufacturer from claiming unlikely levels of performance. Starting in 1920, a series was initiated to examine horsepower, fuel consumption and engine efficiency. There were also practical tests to gauge a tractor's abilities handling implements on a drawbar. These were carried out at the Nebraska State University in Lincoln, and state law stipulated that manufacturers were required to publish all or none of the results, ensuring that they would not extract praise while omitting criticism. The tests were noted for their fairness and authority and led to their general acceptance far beyond the state of Nebraska.

The twenties saw much experimentation with four-wheel-drive tractors as an alternative to crawler machines and Wizard, Topp-Stewart, Nelson and Fitch were among the companies which manufactured them. Ford announced cessation of tractor production at Dearborn in 1928 although it continued briefly in Cork, Ireland before being transferred to Dagenham in 1932. This move was typical of the time and other major corporations were formed from numerous small companies in 1929. Hart-Parr, Nicholas and Shepard and the American Seeding Machine Company all merged with the Oliver Chilled Plow Company to form the Oliver Farm Equipment Company which began to design a completely new line of tractors.

Twin City Tractors, Minnesota Steel and Machinery, the Minneapolis Threshing Machine Company and the Moline Implement Company all merged and Minneapolis-Moline was the result. Thirty-two companies merged to form the United Tractor and Equipment Company which had its headquarters in Chicago,

A 1936 Fordson Model N. This was the first foreign tractor to be tested at the noted Nebraska Tractor Tests and it was submitted for testing in both 1937 and 1938. Ford made 140,000 Model N tractors in England in the war years.

Illinois. Among the 32 was Allis-Chalmers who was tasked with building a new tractor which was known as the United. The corporation did not remain in business long though Allis-Chalmers survived the collapse and continued to build the United tractor, redesignating it the Model U.

Henry Ford demonstrated that there was much common ground between tractor industries on both sides of the Atlantic and a number of small companies were formed around that time. An example is the production of the Glasgow tractor that was built between 1919 and 1924 in the Scottish city of that name. It was a three-wheeled machine, arranged with two wheels at the front and a single driven wheel at the rear to eliminate the need for a differential as did the front wheels which had ratchets in the hubs for the same purpose. The producer of the Glasgow tractor was the DL company which had taken on the lease of a former munitions factory after the Armistice.

Austin of England manufactured a tractor powered by one of its car engines. It sold well despite competition from the Fordson and remained in production into the twenties. In Peterborough another machine, named after its place of manufacture, was made and powered by a complex engine designed by Harry Ricardo while Ruston of Lincoln and Vickers of Newcastle upon Tyne manufactured tractors and Clayton made a crawler tractor. As in America, all were adversely affected by the volume, price and quality of Fordson tractors.

Priorities were different in war-devastated France. The French government made a major effort to regenerate the rural economy and to foster this gave interest-free credit to businesses in the farming sector. Mechanization was vital and tractor trials were instituted at Rocquencourt in the spring of 1920 which tested both domestic and imported models. In the autumn of the same year further trials were held at Chartres and 116 tractors were entered from 46 manufacturers from around the world. French automaker Renault had made light tanks during the war and consequently had proven experience in crawler technology. It turned this application to agricultural machinery and came up with a machine powered by a four-cylinder petrol engine and which utilized tank and commercial vehicle parts. This crawler led to the production of the HO wheeled tractor.

Peugeot offered the T3 crawler and Citroën made a crawler

tractor suited to vineyard use. This latter machine was superseded by the Citroën-Kégresse. The success of tracks off-road led to the development of half-tracks such as the vehicles from Citroën-Kégresse which were proven in expeditions to Africa and the French made the first Sahara desert crossing between December 1922 and January 1923. Later, another French team made the run from Algeria in North Africa south to the Cape of Good Hope between November 1924 and July 1925 while a third expedition in 1931 took French crews from Beirut to the part of French Indo-China now known as Vietnam.

Other French tractor manufacturers included Somua, Amiot, Dubois, Latil, Gerde d'Or, Delahaye, Mistral and RIP and licence-built Saundersons and French-built Austins were also produced. The economic situation meant that, with the devaluation of the French franc, horses remained more commonly used in agriculture than tractors so, with the exception of Latil's timber tractors, Citroën-Kégresse half-tracks, and some Renault machines, interest in tractor production declined.

It was a similar story for the vanquished Germans whose economy was in ruins. At the beginning of the twenties, numerous tractor manufacturers were in business in a small way, although the likes of MAN and Hanomag were more established and produced machines in larger numbers. In 1924 Ford made its presence felt when the Fordson F tractor went on sale in Germany, forcing German manufacturers into competition. The various types of fuel employed by Ford and the Germans illustrated a divergence of ideas where tractors were concerned. German manufacturers, such as Stock and Hanomag, compared the Fordson's fuel consumption unfavourably against their own machines which were moving towards diesel while Lanz introduced the Feldank tractor that was capable of running on poor fuel through use of a semi-diesel engine. The Lanz company later introduced the Bulldog for which the company was notable. The initial Bulldogs were crude; the HL model, for example, had no reverse gear and the engine was stalled and run backwards to enable the machine to be used in reverse. Power came from a single, horizontal-cylinder two-stroke semi-diesel engine that produced 12hp. From this, the HL was gradually improved and became the HR2 in 1926. Lanz was later acquired by John Deere.

The Benz Sendling S7 of 1923 was the first real diesel-engined

tractor and was manufactured by Benz. It featured power from a 30hp two-cylinder vertical engine. The machine itself was a three-wheeled tractor with a single driven rear wheel, although outriggers were supplied to ensure stability during use. A four-wheeled machine, the BK, superseded the S7 and in 1926 Benz and Daimler merged and the Mercedes-Benz engine was used. At around the same time Deutz unveiled the MTZ 222 and the diesel tractor technology race was seriously under way.

Elsewhere in Europe tractor innovation continued apace: in Sweden Avance and its competitor Munktell were offering

improved and redesigned tractors. Avance had considered the starting procedure of diesels in some detail and offered compressed air starters with glow plugs and batteries as options. Munktell offered a similar system. The noted Czechoslovakian motor manufacturers Praga and Skoda offered tractors where their constituent companies had previously offered motor ploughs. A third company, Wilkov, came into the market also and there were now tractor producers in most European countries – Breda, Pavesi, Fiat, Bubba and Landini in Italy, Kommunar in the U.S.S.R., Steyr in Austria, Hofherr and Schrantz in Hungary

33

and Hurlimann and Burer in Switzerland, while further afield there was Ronaldson Tippet in Australia. The concept of the tractor and its mass-production was by now proven and was set to be continually refined, subject to market forces.

Events within the world economy were about to take a turn for the worse with the Wall Street Crash of October 1929. It was described in a headline in *Variety*, the weekly theatrical trade paper, as 'Wall Street Lays an Egg'. The crash led to the Great Depression which continued into the thirties and would see an estimated 13 million Americans out of work by 1932. The United States engaged in several hydro-electric schemes in the late thirties and early forties to provide electricity but also employment in order for the country to work itself out of its grave situation and two dams were constructed at this time, the Hoover and the Grand Coulee. The Hoover, or Boulder Dam, is on the Colorado river between Nevada and Arizona. It stands 736ft (224m) high and 1,224ft (373m) long and began operating in 1936. The Grand Coulee Dam on the Columbia river contains the Roosevelt reservoir and is situated west of Spokane, Washington. It is 550ft (168m) high and 4,173ft (1,272m) long – one of the world's largest hydro-electric plants – and was completed in 1942.

The Wall Street Crash and the economics of production and competition meant that the thirties began on a different note to that of the previous decade. Gone was the optimism and the multitude of small manufacturers with partially proven products. In their place were a few larger companies with fully workable tractors and their new models reflected the increasing use and progress of technology. Simple things, like the oil bath air filter, gave engines a longer life when used in dusty conditions and the pneumatic agricultural tyre was the next step along the path to sophistication. The absence of pneumatic tyres had until the early thirties hampered the widespread 'jack-of-all-trades' use of tractors. Those with lugged wheels could not be used on surfaced public roads while solid tyres suitable for road use were inadequate on wet fields and both solid and lugged wheels caused damage to crops and their roots. The tyre manufacturer B. F. Goodrich tested a zero pressure tyre while Firestone experimented with modified aircraft tyres. These had moulded-on angled lugs and were inflated to around 15 pounds per square inch (psi). Allis-Chalmers was quick to adopt this breakthrough

Harry Ferguson and Henry Ford came to an agreement to produce tractors in Britain and the result was the 9N, virtually a new design. Pictured here is a Ford Ferguson 9N of 1939.

and went as far as hiring racing drivers to demonstrate at speed its new tractors with pneumatic tyres. One driver, Abe Jenkins, used an Allis-Chalmers Model U to break the world tractor speed record with a speed of 66mph (106km/h). Alongside these developments were improvements in vehicle lighting and fuel refining techniques that enabled improvements in tractor efficiency and workability. The Allis-Chalmers WC was introduced in 1934 as the first tractor specifically designed for pneumatic tyres although steel rims were available as an option. To complement these innovations, tractor manufacturers began to be more aware of the importance of styling. At the time, Ford was selling low-priced but well-styled cars such as the V8 Model B while Allis-Chalmers introduced a bright orange paint finish designed to catch the buyer's eye which was a simple ploy but one that no doubt succeeded. Other manufacturers soon followed suit with brightly-coloured paintwork and stylized bonnets, radiator grilles and mudguards.

In Britain, tractor trials were inaugurated in Benson, Oxfordshire in 1930 and the first staging of the event attracted much interest from English and American tractor manufacturers and from makers of the Fordson, at this time being produced in Ireland, while British machines were appearing by AEC, Marshall, Vickers, McLaren and Roadless. It was at the end of the decade that Harry Ferguson and Henry Ford came to an agreement to produce Fordsons in Britain and the result was the 9N. The 9N was virtually a new design (although it had some similarities to the David Brown-manufactured Fergusons of the mid-thirties) and the machines were built in Huddersfield, England.

The Depression had the effect of slowing down innovation rather than eliminating it and did not deter new manufactures from entering the market. In some countries, companies simply had to take their chances in a competitive capitalist market while in others there was less competition. In the former U.S.S.R. – which arose from the Russian Revolution – tractor production continued under the auspices of the state. A Fordson plant was based in the U.S.S.R. but Ford production was halted in 1932 when the factory was switched to the manufacture of a Soviet copy of the Universal, a Farmall model. Elsewhere, in Kharkov and Stalingrad, the International Harvester 15-30 went into

production as the SKhTZ 15-30 and engines based on Caterpillar units were produced at Chelybinsk in the Urals.

The Soviets had a tendency to name their vehicles after the place where the plant was located and later, for example, the Gorky Automobile Works – Gorki'y Automobilni Zavod (GAZ) – in the former U.S.S.R. produced the GAZ-67 from 1942 onwards for the Russian Army and this was superseded in 1952 by the GAZ-69. Production was also undertaken at Ulyanovsk from 1956 and models built there were given the designation UAZ-69. These were light 4x4s, as were the VAZ-2121 models as the Lada Niva is officially designated. The VAZ prefix indicates that it was built at the Volzhky Automobilni Zavod in Togliattrigrad. Nowadays, Belarus tractors are made in the former U.S.S.R. state of Belorussia, now Belarus, and are exported widely.

Elsewhere developments continued apace. In Germany the advent of the diesel engine had changed the face of tractor manufacture. Deutz produced its Stahlschlepper, or Iron Tractor

models, including the F1M 414 and F2M 317 models with single- and three-cylinder diesel engines respectively while Hanomag thrived as a result of exports of both its crawler and wheeled diesel machines. Lanz made advances with its Bulldog models, including the Model T crawler L, N and P wheeled models which produced 15, 23 and 45bhp respectively. They were imported into Britain where the machines were popular because of their ability to run on poor grade fuel, including used engine and gearbox oil thinned with paraffin.

France still trailed the market where tractor production was confined to Renault and Austins made in France. Renault introduced the noted VY in 1933 powered by a 30hp in-line four-cylinder diesel. It had a front-positioned radiator and enclosed engine. It was painted in yellow and grey and became the first diesel tractor to be produced in any significant numbers in France. Other products available included crawlers and specialist machines from such manufactureres as Citroën-Kégresse and Latil as well as smaller vineyard tractors. SFV (Société Française Vierzon) entered the market in 1935 with machines not unlike the Lanz Bulldog and in spite of the Second World War endured until 1960 when it was acquired by Case.

Fiat was still the major producer in Italy and added a crawler to its range in 1932. It was tagged the 700C and was powered by a 30hp four-cylinder engine. Later, but before the outbreak of war, Fiat produced the 708C and Model 40. Breda continued in tractor production and Alfa Romeo experimented with a machine while other makers, such as Landini, Deganello and Orsi, produced Lanz machines under licence. Hurlimann remained in business in Switzerland and produced the 2M20 in 1934 with a two-cylinder engine that produced 20hp. In the former Czechoslovakia, the two-cylinder Wilkov 22 was made by Wichterie and Kovarik of Prostejov while Mavag entered tractor production in Hungary during the thirties.

Further afield in Sweden Bofors, the arms manufacturer, entered the market alongside Avance and Munktell with the 40-46 two-cylinder tractor of 1932. This was the same year that Munktell combined with its engine-maker, Bolinder, to form Bolinder-Munktell and continued the production of the 15-22 and 20-30 tractors. Munktell also experimented with wood-burning tractors, simply because this was a major consideration in

countries where fuel had to be imported. In Australia McDonald Imperial Super diesel tractors were in production although the Depression saw the cessation of Ronaldson Tippet tractor production. In its place came Cliff Howard and the DH22 and saw experimentation with wood-burning and gas tractor power.

Chapter 3
DIG FOR VICTORY 1940–1945

At the outbreak of the Second World War, pronounced in Britain on 3 September 1939 a matter of days after Germany invaded Poland, there were three major tractor producers in business: Fordson, Marshall and David Brown. Fowler's of Leeds was almost immediately commandeered into war work and, to enable farmers to continue to provide food for the nation throughout the oncoming war, it became apparent that tractors would have to be imported from the United States, initially as straightforward purchases and later under the Lend-Lease scheme. Considerable numbers were supplied by Allis-Chalmers, Case, John Deere, Caterpillar, Minneapolis-Moline, Massey-Harris, Oliver, International Harvester and Ford, who also continued production at its factory in Britain, and the machines were subtly redesigned to use less metal and repainted to make them less obviously visible in the countryside.

The situation was significantly altered following the Japanese airstrike on Pearl Harbor, Hawaii on 7 December 1941 which precipitated America into the Second World War. Within days, the United States Marine Corps was fighting a desperate action to hold Wake Island, a tiny Pacific atoll, which had up to that point been used by PanAmerican Airways to refuel its four-engined flying boats for its around the globe services. The island assumed huge strategic importance in the coming struggle for domination of the Pacific, and the Marines' 16-day stand against overwhelming odds becoming the lead story on edition after edition of U.S. newspapers. The *Washington Post* described it as 'the stage for an epic in American military history, one of those gallant stands such as led Texans 105 years ago to cry "Remember the Alamo" ' . It was also the first sign after the disaster of Pearl Harbor that, although the road to victory would be long and costly, America, the most powerful industrial nation on earth, would be the ultimate victor. Preparations were in force to unleash the might of America's industry on the Axis forces of Rome, Berlin and Tokyo. This included America's tractor manufacturers whose products were considered fundamental to army mobility.

In fact, President Roosevelt had declared a limited emergency within a week of the start of the war in Europe on 1 September

RIGHT
White eventually acquired
Minneapolis-Moline, the brand name
of the company being retained by
White until as late as 1969.
Minneapolis-Moline had a long
history and had produced noted
tractors, among them the Universal
and Model U (seen here).

OPPOSITE
This Case DC4 of 1943 was
imported into Britain from the United
States during the Second World
War. It is seen here at the Great
Dorset Steam Fair.

with Germany's invasion of Poland and permitted further recruitment to both the U.S. Army and the National Guard. The process had started that summer when the strength of the army was increased from 175,000 to 210,000 men. General Marshall, recently appointed Chief-of-Staff, established several tactical corps H.Q.s with enough troops to create a fully functioning field army. He also reorganized the basic infantry divisions into five three-regiment 'triangular' Divisions aimed at making them more manoeuvrable and flexible. In 1940 the first Corps manoeuvres since 1918 were held and in May were followed by Corps-v-Corps exercises. While mechanization of the army had begun in 1936 it had been somewhat hampered by lack of funds, although both Indian and Harley-Davidson received contracts for 2,000 motorcycles at a reliability conference called by the Quartermaster Corps at Camp Holabird, Baltimore in November 1938. Now, in 1940, certain expenditure was permitted to purchase much needed transport, including motorcycles. The reason for this was that because of the reorganization it was intended that non-divisional cavalry in the form of cavalry reconnaissance squadrons would ride 'point' ahead of the new divisions. A Squadron would consist of three recce troops and nine recce platoons that would be transported in a defined number of White Scout cars, Dodge Command Cars and motorcycles. It was not, however, motorcycles that would ultimately supply mobility to the army but a light 4x4 machine. Established tractor maker Minneapolis-Moline was among those submitting machines for testing to the U.S. Army under the specification issued by the Quartermaster Corps.

In June 1940 the U.S. Quartermaster Corps issued a specification for a lightweight vehicle capable of carrying men and equipment across rough terrain and invited manufacturers to build prototypes and submit them for testing. Two manufacturers showed sufficient enthusiasm to build prototypes, namely Willys-Overland and American Bantam, both of which were in some financial difficulty at the time. Willys' prototype was late in arriving and Bantam received a contract for 70 vehicles after considerable testing of its machine – the Bantam BRC – at Camp Holabird. Both Ford and Willys were given copies of the blueprints for the Bantam machine which was construed by some as the army expressing its doubts about Bantam's ability to build

OPPOSITE
A large number of the tractors required to enable British farmers to feed the nation throughout the Second World War had to be imported from the United States and American tractor manufacturers supplied their products in considerable numbers. Allis-Chalmers, Case, John Deere, Caterpillar, Minneapolis-Moline, Massey-Harris, Oliver, International Harvester and Ford machines – all were imported into Britain. Pictured here is an example of a Case.

LEFT
The Minneapolis-Moline Model R, seen here in its post-war form, was exported to Europe in large numbers during the war and helped establish the company's standing in Europe.

and supply its 4x4 in the requisite numbers. Further contracts were issued calling for vehicles from Ford and Willys who subsequently submitted their prototypes, the Pygmy and Quad respectively. The machines of all three companies had the early Jeep 'look' and further testing revealed strengths and weaknesses in all of them. Ford submitted a redesigned prototype, the Ford GP, as did Willys – the MA. After further strenuous evaluation tests it was the Willys MA that seemed best overall and in July 1941 Willys was given a contract for 16,000 revised MA models which were referred to as MB. The Bantam BRC and Ford GP quickly faded into the background. Ford, who had a massive manufacturing capability, accepted a contract in November 1941 to manufacture the Willys MB to Willys specifications and the Ford-built examples were known as GPWs. So the legend of the Jeep was born and in every theatre of war, from the mud of the Belgian Ardennes to the jungles of Burma and the sands of Iwo Jima, the Jeep endeared itself to the soldiers of the allied armies. Fighting machine, ambulance, message carrier, mechanical mule, recreational vehicle – the Willys Jeep was them all and more. It was the transport of all ranks from privates to generals in all types of terrain. At the end of the war General Eisenhower was to comment that the Jeep was what had helped America win the war, alongside such machines as the amphibious truck, the Douglas DC3 aeroplane and the Caterpillar bulldozer. By the time the war had ended, Willys had built 358,489 MB Jeeps and Ford 277,896 GPWs.

Current tractor and crawler technology was also turned to military applications and progressed because of it. The U.S. Army had become interested in crawlers and half-tracks when in May 1931 it had acquired a Citroën-Kégresse P17 half-track from France for tests and evaluation. U.S. products soon arrived: James Cunningham & Son produced one in December 1932 and in 1933 the Rock Island Arsenal produced an improved model. Cunningham built a converted Ford truck later in 1933, General Motors became interested and the Linn Manufacturing Company of New York also produced a half-track. In 1936 Marmon-Herrington produced a half-track converted Ford truck for the U.S. Ordnance Department with a driven front axle and towards the end of the decade a half-track designated the T7 made its appearance at the Rock Island Arsenal. It was the forerunner of

This Allis-Chalmers Model B of 1942 was another of the tractors from the big American manufacturers to be exported to Great Britain under the Lend-Lease scheme.

the M2 and M3 models to be subsequently produced by Autocar, Diamond T, International Harvester and White. Half-tracks were to provide the basis for a variety of special vehicles as well as armoured personnel carriers, mortar carriers, self-propelled gun mounts and anti-aircraft gun platforms. The half-track evolved throughout the war years and, although standardized, there were certain differences between models from various manufacturers. Vast numbers were supplied under Lend-Lease to Britain, Canada and Russia and many of the machines produced by International Harvester went abroad under this scheme.

War shortages of such commodities as rubber, as a result of Japanese conquests in the Far East, meant that steel wheels came back into use. Pre-war designs became standardized and remained in production during the war period with only minor and necessary changes being made. In occupied countries, tractor production was mostly halted as their auto industries were turned over to production for the Germans. As the tide of war turned, even the German tractor manufacturers were forced to cease production as a result of bombing and material shortages caused by allied naval blockades. The Germans were desperately short of oil products and farming again became largely reliant on the horse while manufacturers experimented with gas as a fuel. Many tractors were converted to run on gas through use of a conversion kit. Hanomag, Deutz, Lanz Holz, Normag and Fendt all made gas-powered tractors or had their engines converted to use gas. These companies also made numerous wheeled and tracked products for the Wehrmacht.

OPPOSITE
Co-Op tractors were usually the products of other manufacturers such as Cockshutt, for example, sold under the Co-Op name. This is a Co-Op tractor of 1941.

Chapter 4
THE POST-WAR YEARS 1945–1960

The boom in tractor development did not occur until the years following the Second World War. Alongside companies experimenting with and developing the hydraulic bulldozer and loader, others were applying hydraulics to actual excavating machinery. Georges Bataille, a Frenchman, used American techniques developed during the war and introduced his first all-hydraulic design – the TU – in 1950. It was a small machine mounted on a trailer and was powered by the power take-off (PTO) of the tractor towing it. Bataille displayed his machine, named Poclain after a nearby town, at the Agricultural Machine Exhibition in Paris in 1951.

Other pioneers of early hydraulic excavators included Brödr.Söyland A/S of Norway, Blaw-Knox of England and NCK-Rapier, who built the Koehring 505 Skooper. Later companies included the Link-Belt Speeder Division of the American FMC Corporation, Hy-Mac in Britain and Hanomag in Germany.

Once workable designs began to appear they became enormously popular and larger and larger machines were developed with payloads that had previously been regarded as the sole province of cable-operated excavators. So successful is the use of hydraulic systems in this application that there are now countless hydraulic excavators in a variety of sizes built by manufacturers around the world, especially in Europe, the United States and South-East Asia. There are two distinct types of excavators: face or front shovels and backhoes. The major difference between the two is the direction in which the leading edge of the bucket faces. A backhoe draws its bucket back towards itself in order to fill it, while a front shovel pushes it away and choosing the most suitable type of machine for the job depends on the type of excavation work on which it is to be employed. Major manufacturers of excavators were to include Atlas, Case, Caterpillar, Fiat-Hitachi, Hyundai, JCB, and Mannesman Demag.

Light agricultural products were also to prove popular. As the war drew to a close it became apparent that vehicles such as the Jeep would be invaluable to farmers and ranchers so Willys-Overland, who had the foresight to register Jeep as its trade-mark, began to prepare for the production of civilian Jeeps which were

A 16.5hp Oliver 60 row crop tractor of 1947. Oliver's new post-war range included the 66, 77 and 88 models introduced in 1948.

LEFT
A French SFV – Société Française Vierzon – Super 204 tractor made before the company's takeover by Case.

Renault was the major post-war tractor producer in France and had built more than 8,000 tractors by 1948. First came the 303E petrol tractor and then the 3042 diesel from 1948. This is a 1951 Renault tractor.

to be tagged CJs. Initially the Jeep CJs were marketed for agricultural purposes, being equipped with power take-offs and agricultural drawbars. They were heavily promoted as suitable for a variety of farming tasks such as towing ploughs and as disc rotators. The first post-war Jeep was the CJ2A which appeared on the surface to be simply a military Jeep with a different colour finish; but beneath its skin were revised transmission, axles and differential ratios. More obvious was the inclusion of a hinged tailgate and relocation of the spare wheel to the vehicle's side. There were also numerous detail improvements, including bigger headlights and a repositioned gas cap. The engine was, however,

only slightly upgraded from that of the MB. Production of the CJ2A lasted until 1949 by which time there had been 214,202 produced. This production run overlapped with the second of the CJs – the CJ3A. This Jeep went into mass-production in 1948 and continued in production until 1953. The main differences between the CJ2A and CJ3A were a further strengthened transmission and transfer case and a one-piece windshield. In 1953 the CJ3B was introduced with a noticeably different silhouette due to a higher bonnet line which was necessary in order for Willys to fit a new engine. The Hurricane F-head four-cylinder was a taller engine that displaced the same 134 cubic

The first new post-war Fordson tractor was the E27N which was rushed into production at the request of the Ministry of Agriculture. The basis of the machine was an upgraded Fordson N engine with a 3 forward, 1 reverse gearbox with a conventional clutch and rear-axle drive. Production of the E27N was from 1945 to 1951 and in this period various upgrades and options were available. This is a Fordson E27N of 1950 powered by a Perkins diesel engine and seen here at the Great Dorset Steam Fair.

OPPOSITE
The Nuffield Universal was a new British post-war tractor. The Perkins P4 diesel engined variant was introduced in 1950.

LEFT
The 'Grey Fergy' was ubiquitous in British farming but went further afield. Some went to the South Pole with Sir Edmund Hilary's South Pole expedition. This is a 1955 Ferguson TE20 designed for TVO fuel. The TE prefix stands for 'Tractor England'.

inches (2196cc) but produced more horsepower. The CJ3B was to stay in production until the sixties and a total of 155,494 were constructed, destined to live on until the present day through a series of licensing agreements allowing it to be constructed in European, Indian and Japanese factories. Kaiser-Frazer and Willys-Overland merged in 1953 and the resulting Jeep-building company became known as Kaiser-Jeep.

The origins of the Land Rover, on the other side of the Atlantic, are similarly agricultural and it undeniably owes a legacy to the wartime Willys Jeep. Maurice Wilks, chief engineer of the Rover Company bought a war-surplus Jeep for use on his Anglesey estate. In the early post-war years Rover, who had a reputation for producing quality motor cars, was in a difficult

position owing to a shortage of steel and the fact that steel was being allocated to companies producing goods for export, designed to ease Britain's balance of payments. Rover had never seriously exported its cars beyond selling them to Britain's colonies. During the war years, its Coventry factory had been blitzed and it moved out to Solihull where it produced vehicles for the Air Ministry. After investigating all the possibilities of producing a small aluminium-bodied car, Maurice Wilks and his brother Spencer, also a Rover employee, considered the possibility of building a small utility with an aluminium body and four-wheel-drive. The intention was that the machine, specifically intended for agricultural use, would merely be a stopgap until sufficient steel became available for the company to return to building its luxury

A 1953 Massey-Harris Model 745,
seen at work at the Great Dorset
Steam Fair, England.

cars. The Wilks brothers delegated much of the design work to Robert Boyle and a number of employees in the drawing office, and Rover also purchased two war-surplus Jeeps on which to base its design. The designers were limited by other criteria: as far as possible the Land Rover should utilize existing Rover components and, to avoid expensive tooling costs, the panels were to be flat or worked by hand. Unlike steel, aluminium was not subject to rationing which was another advantage. The prototype Land Rovers had a tractor-like centre steering-wheel to make them suitable for either left- or right-hand drive. Because a conventional chassis would have required expensive tooling, engineer Olaf Poppe devised a jig on which four strips of flat steel could be welded together to form a box section chassis. The first prototype featured a Jeep chassis, an existing Rover car rear axle and springs, a Rover car engine and a production saloon gearbox, cleverly mated to a two-speed transfer box and a Jeep-like body. It

ABOVE
A pair of restored Field-Marshalls from the fifties; an orange Series 3A stands alongside a green Series 2.

LEFT
The styling of the pressed steel of the 1957 Series 3A Field-Marshall is distinctive.

A 1953 Renault 7012 with a Perkins diesel engine. Renault entered a similarly powered tractor in the four-day Rambouillet, France tractor trials in October of the following year.

was deemed to have potential and an improved version was given the go-ahead for 50 to be built for further evaluation. A larger, more powerful but still extant, Rover car engine was fitted and the centre steering-wheel was dropped. In almost this form the 80-inch (203-cm) wheelbase vehicle was shown to the public at the 1948 Amsterdam motor show. Orders began to flow in, especially when early Land Rovers were displayed and marketed at agricultural shows around Britain and the company began to look seriously to export markets. An indication of their potential could be gauged from the fact that by October 1948 there were Land Rover dealerships in 68 countries.

The first production models were different from the pilot batch in a number of ways. Some were designed to keep costs down and others to ease production or maintenance. Power take-offs and winches were an extra-cost option and between 1948 and 1954 numerous details were refined and improved. Welders and compressors were mounted aboard Land Rovers and driven by a centre PTO. An early variant was the Station Wagon which consisted of the then traditional wood-framed 'shooting brake' body on a Land Rover chassis, the front wings, radiator grille and bonnet being standard Land Rover panels. While this machine was not the success it might have been, the idea was to be recycled more successfully in later years.

The more established tractor manufacturers, such as the nine largest American companies, were quick to add new models to their ranges. Allis-Chalmers, for example, introduced the Models G and WD. John Deere replaced the Models H, LA and L with the Model M in 1947. This was followed in 1949 by two derivatives, the Model MC and MT – a crawler and tricycle – and this was followed with the diesel Model R. International Harvester returned to tractor manufacture and replaced the A with the Super A that had a hydraulic rear-lift attachment as hydraulics began to be more widely utilized.

Massey-Harris brought out a new range in 1947 which included the Model 44 but also the 11, 20, 30 and 55 models. Oliver followed a year later with a new range, diesel options including the 77 and 88 models: otherwise the tractors used the four-cylinder petrol and paraffin engines. PTO equipment was standard but hydraulic lifts were not. Ford and Ferguson went their separate ways when the latter failed to concur with the

younger Henry Ford's ideas regarding the future of tractors. This split led to a certain amount of litigation and saw the opening of Ferguson's own Detroit factory. The upshot was that Ferguson won almost $10 million worth of damages for patent infringement and loss of business. Another feature of the post-war years was the influx of new manufacturers which included Brockway, Custom, Earthmaster, Farmaster, Friday, General, Harris and Laughlin.

In Britain the first new post-war tractor was the Fordson E27N and it was rushed into production at the request of the Ministry of Agriculture. The basis of the machine was an upgraded Fordson N engine with a 3 forward, 1 reverse gearbox with a conventional clutch and bull wheel and pinion rear-axle drive. The production span was from 1945 to 1951 and various upgrades and options became available in that period. These

Post-war, Massey-Harris introduced the Model 44 based around the same engine as the pre-war Challenger and using a five-speed transmission. The Model 30 was introduced in 1947 with a five-speed transmission and more than 32,000 were made before 1953. That was the year this 745 was made, the company also merging with Ferguson.

included electrics, hydraulics and a diesel-engined version. Ferguson came to an arrangement with Sir John Black of the Standard Motor Company to produce a new tractor in a Standard factory. It utilized an imported Continental engine and production began in 1946. Production continued with Standard's new engine which became available in 1947 and a diesel option appeared in 1951. This tractor – the TE20 – was nicknamed the 'Grey Fergy' – a reference to its designer and its drab paintwork. It became enormously popular and ubiquitous on British farms.

J. Wentworth Day, writing on farming topics in *The New Yeomen of England*, in 1952, said of one farm: 'Three Ferguson Tractors are in use, and this handy, adaptable and powerful tractor seems likely to become a general maid-of-all-work.' Writing at the same time of a Christian community farm he remarked: 'The Society of Brothers has standardized on one make of tractor – the Ferguson with its range of implements – and find the Ferguson System ideally suited to hill farming. Not only have the tractors handled all the ploughing and other work with cultivators, mower, earth-scoop, Paterson buck rake, etc., but equally importantly have eased estate management problems by reason of their dependability.'

An unusual application of Fergusons is described by Sir Edmund Hilary and Vivian Fuchs in their book entitled *The Crossing of Antarctica* which contains numerous references to the tractors used on the Commonwealth Trans-Antarctic Expedition. These tractors were used to tow sledges, unload ships, and for reconnaissance, and the tracked Sno-Cats and Studebaker Weasels were also employed. The Fergusons were for at least part of the time equipped with rubber crawler-type tracks around the front and rear wheels with an idler wheel positioned between the axles. The expedition was glad of the Fergusons' abilities on numerous occasions and one was driven to the South Pole itself. As pointed out in the book: 'Certainly our Ferguson tractors in their modified and battered condition did not inspire confidence in the casual observer but we still felt that their reliability and ease of maintenance would counteract to some extent their inability to perform in soft snow like a Sno-Cat or a Weasel.'

The Yorkshire-based David Brown company, having made airfield aeroplane tractors during the war years, reintroduced its VAK 1 as the slightly improved VAK 1A until it was able to introduce the Cropmaster of 1947. This was a popular machine, especially in diesel form as introduced in 1949, while Nuffield entered the tractor market with the Universal. Three of the large American manufacturers set up factories in Britain: IHC in Doncaster, South Yorkshire; Allis-Chalmers in Hampshire; and Massey-Harris in Manchester, although it transferred to Kilmarnock, Scotland in 1949. IHC was the only one of the three to have any success. Renault became the major force in the post-war tractor industry in France and managed to build more than 8,500 tractors by 1948, many of which included the 303E model although it followed this with the 3042.

West Germany, occupied by the Allies, began to assume a semblance of normality. There are records of elementary light tractors being made from damaged war-surplus Jeeps and recycling of this nature also took place in Britain and the Philippines. Established companies like Lanz and Hanomag resumed full production once their factories were returned to normal while MAN produced some 4x4 tractors which, with the

MAN 325, were the forerunners of the Mercedes-Benz Unimog. New companies were founded, including Alpenland, Normag, Primus, Stihl and Faun. Of the latter, Faun went on to make lorries and Stihl chain-saws. Fendt returned to tractor production, as did Deutz, and the German International Harvester factory. This had been badly bomb-damaged; but once the task of reconstruction was completed in 1948 it produced the FG12, a version of the F12 Farmall. A restyled diesel variant followed in 1951.

Italian tractor production in the immediate post-war years was, in the main, a continuation of the pre-war Lanz variants while the country, partially destroyed by fighting, rehabilitated itself. In Austria, Steyr resumed tractor production and much further away in Australia the Chamberlain 40 went into production. This machine was built in a war-surplus factory and was powered by a horizontally-opposed 30hp engine. Kelly-Lewis manufactured a basic machine, loosely based on a pre-war Lanz, but the model was soon superseded.

Developments also occurred in the countries that now comprised the Eastern Bloc. The imposition of Communist rule led to the industries of these nations being completely reorganized, even where tractor companies already existed. As a result, Skoda, Zetor, HSCS, IAR, Ursus and Aktivist came into being or were reorganized in Czechoslovakia, Hungary, Romania, Poland and the DDR respectively. In the U.S.S.R. the Universal went back into production and the new Stalinetz 80 was a Communist-designed crawler tractor.

The fifties was a time of prosperity for tractor manufacturers in a decade that brought both a booming economy and a period of relative stability to bear on agriculture. Tractor manufacturers had proliferated in the post-war years to the extent that there were now 45 manufacturers in the United States, 20 in Britain and as many as 60 in Germany. Diesel tractors were becoming more popular in the United States and the three-point lift became the norm across the industry. The respected Nebraska Tractor Tests were revived towards the end of the decade although the practical part of the tests now focused on PTO applications rather than belt drives. Massey-Harris and Ferguson combined in 1953 to form what became the Massey-Ferguson Company, in 1958.

LEFT
Ronald Hoof's 1950 Turner Yeoman Model 2A tractor powered by a 40hp V4 engine.

OPPOSITE
Case introduced a new series of tractors in the early fifties. Among them were the 300, 400 and 500 series. This 1952 Case, a Model 500, is one of the latter.

A competitor in the European ploughing championships using a diesel David Brown 30D made in Huddersfield, England.

RIGHT
During the fifties, the countries of the
Eastern Bloc, including East
Germany, Czechoslovakia and
Poland, produced a standardized
tractor in the ZT, Zetor and Ursus
factories respectively. This is an
Ursus.

OPPOSITE
A restored UTB Universal 530 tractor
at an agricultural show.

Chapter 5
THE MANUFACTURERS

OPPOSITE
A 1997 AGCOStar 8425 articulated four-wheel-drive tractor. This machine was introduced in 1995 and customers had a choice of either an 855ci Cummins or a 744ci Detroit diesel engine. Both produced 425hp.

AGCOStar

This new brand, introduced by AGCO, appeared in 1995. The models 8350 and 8425 were state-of-the-art machines that used the latest technology to provide a tractor suitable for large acreage farming. The machines are based around a channel chassis that is articulated in the centre to enhance turning. Both axles are driven and there is an 18 forward, 2 reverse gearbox.

Specification

Make	AGCOStar
Year	1997
Model	8425
Engine	Detroit diesel
Power	425hp
Notes	18F, 2R transmission, centre-articulating

Allis-Chalmers

This famous company dates back to the turn of the century, being formed in 1901 by the merger of four other companies. The new company built its first tractor in 1914, the Model 10-18. The Model 18-30 tractor was introduced in 1919 and over the course of the next decade approximately 16,000 were made. Also introduced in 1919 was the Model 6-12. Allis-Chalmers acquired a few other companies during the thirties including the Advance Rumely Thresher Company.

The Model U and E tractors formed the basis of the Allis-Chalmers range and in the late thirties its Model A and B tractors were introduced. The four-speed Model A replaced the Model E and was made between 1936 and 1942 while the Model B ran between 1937 and 1957. The Model B was powered by a four-cylinder 15.7bhp engine and more than 127,000 were made. In 1936 the Model U was upgraded when it was fitted with the company's own UM engine. The WC with pneumatic tyres was introduced in 1934 as the first tractor designed for pneumatic tyres, although steel rims were available as an option. In 1938

In 1990, the German-owned company that had controlled Allis-Chalmers was acquired by an American holding Company known as the Allis Gleaner Company (AGCO). The company was renamed AGCO-Allis and produced machines such as this AGCO Allis 9630.

Allis offered the down-sized Model B tractor on tyres and this was a successful move. The tractor was a sales winner and was widely marketed, manufactured in Britain after the Second World War for sale in both the British and export markets. In this year the company increased the number of its tractor models with styled bonnets and radiator grille shells.

In the post-war years Allis-Chalmers introduced the WD-45 model in both petrol- and diesel-engined forms. The WD-45 was a great success and more than 83,000 were made. In 1955 Allis-Chalmers bought the Gleaner Harvester Corporation and the company introduced its D Series in the latter part of the decade. This was a comprehensive range of tractors that included the D-10, D-12, D-14, D-15, D-17, D-19 and D-21 machines.

The final Allis-Chalmers tractor was produced in the mid-eighties when in 1985 the company was taken over by a West

German company. Klockner-Humboldt-Deutz AG took control and renamed the tractor division Deutz-Allis. This was a short-lived liaison and in 1990 the company was acquired by an American holding company known as the Allis Gleaner Company (AGCO) when the company was renamed AGCO Allis.

Specification

Make	Allis-Chalmers (U.S.A.)
Year	1986
Model	4W-305
Engine	Twin turbo
Power	305hp
Notes	20F, 4R transmission

An Allis-Chalmers 8050 from the early eighties. It is powered by a 426ci (6980cc) turbo-charged six-cylinder engine.

A 1997 Belarus 1005. Belarus tractors are made in Belarus, now part of the Commonwealth of Independent States (formerly the U.S.S.R. Belorussia), and are exported widely.

Belarus

Belarus Tractors are made in the former U.S.S.R. state of Belorussia, now Belarus – a member of the CIS (Commonwealth of Independent States) – and has exported its tractors widely. It offers numerous tractors in a variety of configurations, the various models ranging from 31- to 180-bhp machines. The Belarus 925 features 100bhp while the 1770 has a 180-bhp engine.

Specification

Make	Belarus
Year	1986
Model	862 D
Engine	249ci (4075cc)
Power	90hp
Notes	Automatic 4x4 transmission; 18F, 2R gears; hydrostatic steering

Case

The J. I. Case Threshing Machine Company was formed in 1863 to build steam engines. Its first tractor appeared in 1892 and the company went on to become one of the leaders of the industry having as early as 1923 produced 100,000 tractors. In 1929 the Model L went into production based around a unit frame construction. Case assumed control of David Brown tractors in 1972.

Case and International Harvester merged in 1984 to form Case-IH and tractors built since this date have incorporated ideas from both companies

Specification

Make	Case-IH
Year	1986
Model	685L
Engine	4-cylinder
Power	69hp
Notes	8F, 2R 4x4 transmission; hydrostatic steering

Specification

Make	Case-IH
Year	1986
Model	1594
Engine	6-cylinder
Power	96hp
Notes	Four range semi-automatic Hydra-Shift transmission

A 1996 Case 4230 4WD tractor in operation. Modern tractors such as this have cabs insulated to cut down on vibration.

A David Brown 995, equipped with a hydraulic front loader, at work on the harvest.

David Brown

The Ferguson tractors of the mid-thirties, manufactured by David Brown, came about when Harry Ferguson approached the company regarding a proposed tractor transmission. Brown's was noted for making gears and Ferguson wished to produce a tractor with an American Hercules engine and an innovative hydraulic lift. These machines were to be built in Huddersfield, England by Brown after the prototype had been tested. It was called the Ferguson Model and was fitted with a Coventry Climax engine and latterly an engine of Brown's own design. Production ceased in 1939 because Brown wished to increase power and Ferguson to reduce costs. Ferguson went to the United States to see Ford while David Brown exhibited a new model of tractor to his own design. The new machine was the VAK 1 and featured a hydraulic lift. After the war, David Brown reintroduced its VAK 1 as the slightly improved VAK 1A until it was able to introduce the Cropmaster of 1947. This became a popular machine, especially in its diesel form which was introduced in 1949. The company was taken over by Case in 1972.

A Ferguson T20 with a four-wheel-drive conversion to the front axle by Selene of Turin, Italy.

Ferguson

Harry Ferguson's liaison with the David Brown company came to an end when production of the Ferguson Model (*see opposite*) ceased in 1939. He decided to make the trip to the United States for talks with Henry Ford which resulted in his three-point system being introduced on Ford's 9N model. This system allowed for the attachment of various farm implements and was produced in co-operation with the Ferguson-Sherman Company until 1946. Ford and Ferguson went their separate ways when Ferguson disagreed with the younger Henry Ford's views on the future of the tractor industry. The split led to litigation and saw the inauguration of Ferguson's own Detroit factory. The upshot was that Ferguson won from Ford almost $10 million damages for patent infringement and loss of business.

Harry Ferguson later came to an arrangement with Sir John Black of the Standard Motor Co. to produce a new tractor in a Standard-owned factory. Production of the TE20 began in 1946 and Standard's engine was used in 1947 and production continued with the new engine when a diesel option was made available in 1951. Ferguson later sold his company to Massey-Harris.

LEFT
The Fiat Agri range, which includes this F140 from 1994, is powered by a water-cooled turbo-charged in-line six-cylinder engine.

Fordson

Alongside its cars and trucks, for which it is probably more famous, Ford has always been a major producer of tractors. Henry Ford built his first tractor in 1915 and by 1916 had a number of working prototypes under evaluation. His aim, in producing an affordable tractor, was to do for farmers what the Model T had done for motorists. With help from his staff Ford had designed what would later become the Fordson F model. Its secret was that it was of a stressed cast-iron frame construction. This frame contained all the moving parts in dust-proof and oil-tight units which eliminated many of the early tractors' weaknesses. Ford's prototypes, seen at a tractor trial, were considered to be a practical proposition and the government requested that they be put into production immediately. He was preparing to do exactly that when, in the early summer of 1917, German bombers attacked London in daylight. The British government recognized the seriousness of the threat and proposed to turn all available industrial production over to the manufacture of fighter aeroplanes for the defence of Britain.

It was requested that Ford should make his tractors in the United States, which was agreed, and only four months later F models were being produced. While this was only a temporary setback, the delay and shift in production caused Ford another problem. His plans and intentions became public knowledge before the tractors actually appeared and another company attempted to pre-empt him in the meantime. This Minneapolis concern formed the Ford Tractor Company using the name of one of its engineers and by doing so deprived Henry Ford of the right to call his tractors Ford. He resorted to the next best thing – Henry Ford and Son – shortened to Fordson. The rival 'Ford' tractors produced by the Ford Tractor Company failed to amount to very much. It managed to sell one to a member of the Nebraskan establishment who was less than satisfied with it when he compared it to a Bull tractor which he had also acquired. This led to the establishment of the Nebraska Tractor Tests which became respected worldwide.

The British Army ordered 5,000 of these Fordson tractors for the war effort and production of them in the United States ceased in 1928; but elsewhere, including Ireland and later England, it continued. Production recommenced in the United States during the forties and has continued ever since. Fordson made the Model F in the twenties and sold the model in vast numbers. Ford produced tractors that were reliable and gradually incorporated refinements as technology advanced. As early as 1918 Fordsons had a high-tension magneto, a water pump and an electric starter.

More important, though, was the introduction of the three-point system in the late thirties. It was introduced on Ford's 9N model having been designed by Harry Ferguson, an Irishman. This system allowed for the attachment of a variety of farm implements and was produced in co-operation with the Ferguson-Sherman Company until 1946. Ford's British factory had been at Dagenham since 1932 and from there Fordsons were exported around the world as well as back to the United States. This included the All Around row crop tractor built there specifically for the United States in 1936. The Fordson was the first foreign tractor tested at the noted Nebraska Tractor Tests and was submitted for examination in both 1937 and 1938.

Another Fordson plant was based in the U.S.S.R. where production was halted in 1932 when the factory switched to making a Soviet copy of the Universal Farmall.

Ford made 140,000 Model N tractors in England in the war years. Its first new post-war tractor was the Fordson E27N and it was rushed into production at the request of the Ministry of Agriculture. The basis of the machine was an upgraded Fordson N engine with a 3 forward, 1 reverse gearbox with a conventional clutch and rear-axle drive. Production of the E27N was from 1945 to 1951 and various upgrades and options became available in this period. These included electrics, hydraulics and a diesel-engined version. Ford's 50th Anniversary was in 1953 and in that year the Model NAA was introduced.

By the sixties Ford was producing tractors in Brazil, England, India and the United States. The Model 8000 was the first Ford tractor to have a 100-hp engine, displacing 401ci (6571cc).

In 1986 Ford bought the New Holland Company and in 1987 the Versatile Tractor Company of Canada. British production of Ford tractors is carried out at Basildon in Essex, one of Ford's eight manufacturing plants around the world, the company exporting to 75 countries. The 10 series consists of 11 tractors ranging from the 177ci (2900cc) 44-hp 2910 model to the 403ci (6600cc) 115-hp 8210 model.

A Ford 6810. The recession of the eighties, and the contracting tractor market world-wide, led to the merger of tractor operations of both Ford and Fiat under the New Holland brand-name.

Specification

Make	Ford (U.K.)
Year	1986
Model	4610
Engine	403ci (3300cc)
Power	64hp
Notes	One of 11 tractors from the '10' series

Specification

Make	Ford (U.K.)
Year	1986
Model	7610
Engine	268ci (4400cc)
Power	103hp
Notes	One of 11 tractors from the '10' series

RIGHT
An Italian-made 1997 Landini
Advantage 65F.

OPPOSITE
The Landini Legend 130 is in the
company's 1997 range and is larger
in capacity than the 65F model.

International Harvester exported tractors to Britain where it marketed the models from its range that were considered to be most suited to British conditions. These were the Titan 10-20 and the Mogul 8-16. This is a 1919 International Harvester Titan 10-20.

International Harvester

The origin of International Harvester, and probably of its 'Cornbinder' nickname, goes right back to 1831. In that year Cyrus McCormick invented a reaper which, while officially referred to as the McCormick Reaper, soon became known as the Cornbinder. The company became known as the International Harvester Corporation in 1902 and by 1907 was producing Auto-wagons, a very early light-duty truck. Things went well for the company and by 1912 buses and more substantial trucks were being produced. The company really took off after the First World War when, in 1921, a line of new trucks designated the S series was put on sale. They were sold through farm equipment dealers and became immediately popular with farmers. They were painted a distinctive shade of red and because of this and their small size became known as 'Red Babies'. The 'S' stood for Speed-truck as the Red Baby was capable of 30mph (48km/h).

Life was good for IHC throughout the twenties; in 1929, for example, it sold 50,000 trucks through 170 outlets. This level of production put it in the big league with companies such as Ford and Chevrolet. It entered the thirties optimistically, a landmark being a restyle introduced in 1934 that gave the trucks car-like styling. This was superseded by another new design in 1937 and was in turn replaced by the K-series introduced in 1940. Post-war came the KB model and the Travelall, a station wagon-type panel truck. Then in the fifties came the L, R and S series trucks.

International's truck production was of course carried out alongside its tractor production. Three models carried the company through the twenties: the 10-20, 15-30 and Farmall Regular and these were refined and redesigned for the thirties. The Farmall Regular became one of a range of three models as the F-20, with the F-12 and F-30. These machines were similar but had different capacities. In 1929 the 15-30 had become the 22-36 and was subsequently replaced by the W-30 in 1934. The 10-20 had a long production run which continued until 1939. IHC dropped its grey paint scheme during the thirties, replacing it with red. The effect of this was two-fold: firstly, it made tractors more visible to other motorists as they began to appear on public roads and had the secondary effect of being more noticeable to potential buyers. The first IHC wheeled, diesel-powered tractor appeared in 1934, designated the WD-40. It was powered by a four-cylinder engine and, as indicated by its numerical suffix, had in the region of 44bhp. The engine was slightly unorthodox in that it was started up on petrol and, once warm, switched to diesel through closure of a valve. During the Second World War IHC manufactured a range of machinery, including half-track vehicles for the allied armies. The company produced more than 13,000 International M-5 half-tracks at its Springfield plant. The company also made a quantity of 'essential use' pickups for civilians who required transport in order to assist the war effort. Civilian production fully resumed in 1946 with the reintroduction of the K-series.

The first IH tractors appeared in 1906 when the Type A gasoline tractor was marketed with a choice of 12-, 15- and 20-hp engines. The Type B soon followed and lasted until 1916. IH was noted for its production of the giant 45-hp Mogul tractor during the second decade of the 20th century. It followed this in 1919 with the Titan, a 22-hp machine. Farmall was one of the trade-names used by IH during the twenties. The name endured for several decades and machines such as the Farmall M of the fifties is fondly remembered. By the seventies, the Farmall 966 was a 100-hp machine with a 16 forward, 2 reverse gear transmission and a 414ci (6784cc) engine.

JCB

The famous British plant manufacturer JCB produced a comprehensive range of machines including the famous 'digger' with a face shovel and back hoe. It also produced wheeled and tracked excavators and within this range are eight different tracked crawler excavators. These range from the JS70, a compact machine designed especially for use in confined spaces, with an operating weight of 7 tons, to the JS450LC with an operating weight of 44 tons. In between is the JS200 which is produced in two versions, the JS200/220 and the JS220LC Long Reach. The latter has an extremely long boom and dipper arm to make it suitable for specialist applications such as waterway maintenance. To enable it to carry out such tasks, specialist buckets are available, including one for weed mowing.

The JCB 3CX-4 is one of many variants of the long-running digger from J. C. Bamforth. Although the JCB in this guise is designed specifically as an excavator, with front loader and back hoe, it is clearly tractor-derived.

John Deere

John Deere is the last tractor manufacturer to have its founder's full name as its brand-name. The John Deere colours of green with yellow detailing remain too, as vibrant as ever. The company began the manufacture of horse-drawn steel ploughs in 1836 and chose not to involve itself in the production of steam-powered vehicles but did make a move to internal combustion-engined tractors in the early 1900s. During the twenties, the company introduced its two-cylinder Model D tractor. This was a particularly successful machine and remained in production, albeit upgraded, until the fifties. The Model A was another machine produced between 1934 and 1952. It was acknowledged to be the most popular tractor ever made by John Deere and incorporated some innovative ideas. The wheel treads were adjustable through the use of splined hubs and the transmission was contained in a single-piece casting. The first Model A was rated at 24bhp but this output was subsequently sequentially increased. John Deere's Model B was made between 1935 and 1952 in a variety of forms, including the model BO of the forties and the crawler track-equipped version in the late forties. The later MC model was purpose-built as a crawler tractor.

John Deere's first diesel tractor was produced in 1949 and was tagged the model R before the company switched to numerical designations. The machines that followed were the 20 series of the mid-fifties, the 30 series introduced in 1958 and the 40, 50 , 60, 70 and 80 series that followed through the fifties and sixties, the 50 series, for example, replacing the Model B in 1952.

The John Deere company introduced its first 4x4 tractor in 1959. During the seventies the 40 series tractors were popular as by now they incorporated V6 turbo-charged diesel engines that produced in excess of 100bhp. This latter figure was exceeded by the end of the next decade when John Deere's 4955-model tractor produced 200bhp. The 7800 has dual rear wheels at each side and a 466-ci (7636-cc) engine that produces 170bhp and drives through a 20-speed transmission.

Although John Deere is primarily an American tractor manufacturer, it also has production plants in other countries including Argentina, Australia, Germany, Mexico, South Africa and Spain.

Specification

Make	John Deere (U.S.A.)
Year	1986
Model	4450
Engine	464ci (7600cc)
Power	160hp
Notes	15-speed transmission, part-time 4x4

Specification

Make	John Deere (U.S.A.)
Year	1986
Model	2040
Engine	238ci (3900cc) diesel
Power	70hp
Notes	One of a series of 3 tractors, 1640, 2040 and 2140 models

Current John Deere tractors, such as this 7600 FWD model, have up to 20 speed transmissions and six-cylinder engines.

The company is Canada-based and produces tractors all around the world. Massey-Ferguson itself became part of AGCO in 1994. The first post-merger tractor was the MF-35 of 1960 and this was soon followed by a range of tractors, including the MF-50, MF-65 and MF-85.

In 1976 Massey-Ferguson introduced the 1505 and 1805 machines. These both featured a 174-hp Caterpillar V8 diesel engine. In the mid-eighties the 3000 series of Massey-Ferguson machines was made available with a turbo-diesel 190hp six-cylinder engine. Another important Massey-Ferguson tractor was the MF 398, powered by a 236-ci (3867-cc) diesel engine and featuring 4x4 transmission.

Massey-Harris

Massey-Harris was formed in 1891 in Toronto, Canada from the merger of two competing manufacturers of farm implements. Daniel Massey had been making implements since 1847 while A. Harris, Son & Company was in competition for the same market. The Model 25 was a popular tractor through the thirties and forties. Another Massey-Harris tractor of this era was the 101 of 1935. This machine was driven by a 24hp Chrysler in-line six-cylinder engine. A third was the Twin Power Challenger powered by an I-head four-cylinder engine that produced approximately 36hp. The company chose a red-and-straw yellow colour scheme for its tractors in the mid-thirties.

Post-war, Massey-Harris introduced the Model 44 based around the same engine as the pre-war Challenger and using a five-speed transmission. The Model 30 was introduced in 1947 with a five-speed transmission and more than 32,000 were made before 1953 when the company merged with Ferguson (*see above*).

ABOVE
In the mid-eighties, the 3000 series of Massey-Ferguson machines was made available with turbo-diesel 190hp six-cylinder engines. The Massey-Ferguson 3115 seen here is one of this range.

OPPOSITE
The post-war Massey-Harris line was introduced in 1947: styling was similar to earlier models and the range encompassed the Model 11 Pony (*seen here*) and the much larger Massey-Harris 55.

Massey-Ferguson

Harry Ferguson sold his tractor company to Massey-Harris in 1953. The latter company had been producing tractors in an attempt to compete with both Ford and Ferguson. Once the purchase of Ferguson was completed, the company offered the MH50 and Ferguson 40. These had different bodyworks but were mechanically identical. Both of these models were based on the Ferguson 35 that had come as part of the buy-out deal. In 1958 the company was renamed Massey-Ferguson and another new line was introduced. The Massey-Ferguson 65 and 35 were the new models and were powered by Perkins diesels, Massey-Ferguson acquiring Perkins in 1959.

Specification	
Make	Massey-Ferguson
Year	1986
Model	2685
Engine	354-ci (5800-cc) Perkins
Power	142hp
Notes	One of the 2005 series of three tractors, 2645, 2685 and 2725; 4x4, 16F, 12R gears; turbo-diesel

Specification	
Make	Massey-Ferguson
Year	1986
Model	MF699
Engine	354ci (5800cc)
Power	100hp
Notes	Most powerful of the MF600 series of four tractors, MF675, 690, 698T and 699

Mercedes-Benz

Mercedes-Benz of Germany has been producing Unimogs for several decades and they can be counted among some of the most capable off-highway machines ever manufactured. The 6x6 U2450 L model is one of the largest with a permissible gross vehicle weight of 37,478lb (17,000kg). The Unimog features portal axles with disc brakes, an eight-speed manual transmission and power-steering. The range of Unimogs is fully comprehensive and includes numerous 4x4 versions.

Specification

Make	Mercedes-Benz
Year	1996
Model	U2450
Engine	Mercedes-Benz OM 366 LA
Power	177kW @ 2600rpm
Notes	8F, 2R gears, 6x6 transmission

Specification

Make	Mercedes-Benz (Germany)
Year	1986
Model	MB-trac 1500
Engine	346ci (5675cc)
Power	150hp
Notes	Dual-directional tractor, turbo-diesel engine

Roadless

Roadless Traction Limited of Hounslow, England designed a number of half-track bogie conversions for vehicles as diverse as a Foden steam lorry and a Morris Commercial. It later devised tracked conversions for various makes of tractors, including the Fordson E27N, as well as forestry tractor conversions for Land Rovers.

Volvo BM

Volvo BM is one of the world's largest manufacturers of tractors, articulated haulers, wheeled loaders, rigid haulers and hydraulic excavators. The company's products are marketed all around the world under the well known names of Volvo BM, Valmet, Michigan, Euclid, Zettelmeyer and Akerman. Volvo BM offers a comprehensive range of tractors.

Specification

Make	Volvo BM Valmet (Sweden)
Year	1986
Model	2105
Engine	6-cylinder turbo-diesel
Power	163hp
Notes	Intercooled turbo-charged diesel engine

A Volvo BM – Bolinder-Munktell – 650 Turbo in rural France.

RIGHT
A Renault 180-94 turbo-diesel-
powered tractor from 1996.

OPPOSITE
The Steiger Panther 1360 tractor is
a four-wheel-drive machine, powered
by a large displacement Caterpillar
diesel engine.

The White 6195-PFA is one of a range of tractors that includes the 6125-PFA, 6175-2WD, and 6215-PFA models. The output of these machines ranges from 125 to 215hp.

White

White became established in the tractor-manufacturing industry during the sixties when the corporation bought a number of smaller tractor-producing companies. These smaller companies were in the main well established concerns with a long history of involvement in the tractor industry stretching back to its earliest days. In 1960 White bought Oliver and continued to produce the successful and popular Oliver 55 models throughout the decade. Oliver had been in business since 1929, also formed as a result of combining several small companies when the Oliver Chilled Plow

Specification

Make	Oliver (U.S.A.)
Year	1960
Model	55
Engine	Continental
Power	49hp
Notes	Production of the Oliver 55 continued after Oliver was taken over by White

Works, Hart-Parr, Nicholls and Shepard and the American Seeding Machine Company came together under the Oliver brand.

During the thirties, Oliver manufactured such important tractors as the Oliver 90 which was a three-speed 40 hp machine while the Oliver 66 and 77 Row Crop models were stalwarts of the company line-up between the late forties and the sixties. The Super 99 was a three-cylinder two-stroke supercharged General Motors diesel-powered tractor with a displacement of 230ci (3769cc). It was equipped with a three-point hitch and a cab was an extra cost option.

White then purchased the Cockshutt Farm Equipment Company in 1962, a company whose history stretched back to 1839. In more recent times Cockshutt had manufactured tractors such as the Models 20 and 40. The 20 was introduced in 1952 and utilized a Continental L-head 140ci (2294cc) four-cylinder engine and a four-speed transmission. The Model 40 was a six-cylinder, six-speed tractor.

Following this, White acquired Minneapolis-Moline a year later, the brand names of all three companies being retained by White until as late as 1969. Minneapolis-Moline also had a long history and had produced noted tractors such as the Universal and Model U. The 27-hp Universal was made between 1915 and 1923. In 1987 White merged with the New Idea Farm Equipment Company and subsequently was itself acquired. In 1991 the AGCO Corporation of Waycross, Georgia purchased White although it has retained the White name for its brand of tractors. During the seventies White produced a tractor named the Plainsman. This was an eight-wheeler, powered by a 504-ci (8259-cc) engine that produced 169hp.

Zetor

Zetor is a Czech Republic company and versions of its tractors are assembled in numerous countries around the world, including Argentina, Burma, India, Iraq, Uruguay and Zaire.

The Zetor 7540 is a four-wheel-drive tractor powered by a four-cylinder engine of 71hp. Zetor tractors are manufactured at numerous plants around the world.

Specification

Make	Zetor (Czech Republic)
Year	1986
Model	8045 Crystal
Engine	278ci (4562cc) diesel
Power	85hp
Notes	4x4 with 8F, 2R gears

Chapter 6
CRAWLER TRACTORS

Running parallel to the development of the steam excavator were experiments with tracked machines known as 'crawlers'. These first experiments involved wheeled steam tractors which were converted to run with tracks. The first test of such a conversion, to a Holt steam tractor, took place in November 1904 in Stockton, California. This had been accomplished by removal of the wheels and the rears' replacement with tracks made from a series of 3x4-inch (7.6x10-cm) wooden blocks bolted to a linked steel chain which ran around smaller wheels, a driven sprocket and idler on each side. Originally the machine was steered by a single tiller-wheel although this system was later dropped in favour of the idea of disengaging drive to one track by means of a clutch which slewed the machine around. From there it was but a short step to gasoline-powered crawlers, one of which was constructed by Holt in 1906. By 1908, 100 such crawlers were engaged in work on the Los Angeles aqueduct project in the Tehachapi mountains.

In the years following the First World War, the Best Company continued its work with crawler tracked machinery and in 1921

introduced a new machine, the Best 30 Tracklayer. This crawler was fitted with a light-duty bulldozer blade powered by an internal combustion engine and had an enclosed cab. In the early days there had been a considerable amount of litigation involving patents and types of tracklayers and two companies were frequently named in disputes: Best and Holt. Holt's patent for tracklayers enabled him to charge a licence fee to other manufacturers of the time, including Monarch, Bates and Cletrac. Then the First World War intervened and much of Holt's production went to the U.S. Army while Best continued to supply farmers. The two companies later competed in all markets and neither had a significant advantage over the other. Eventually, in 1925, Holt and Best effectively merged to form the Caterpillar Tractor Company and the new company, later that year, published prices for its product line: the Model 60 sold for $6050, the Model 30 for $3665 and the two-ton $1975. The consolidation of the two brands into one company proved successful in the next few years, the prices of the big tracklayers were cut, business increased and sales more than doubled. Starting in 1931, all

Long-wheelbased Land Rovers were available for a time with crawler conversions made by Cuthbertson and Sons of Biggar, Scotland. The conversion was a subframe that bolted to the Land Rover chassis with eight bogie wheels. Each crawler track was driven by a sprocket that bolted to the wheel hub of the Land Rover in place of the standard wheel and tyre.

Caterpillar machinery left the factory painted Highway Yellow which was seen not only as a way of brightening up the machines in an attempt to lighten the gloom of the Depression but also as a way of ensuring the safety of machines increasingly being used in road construction, which had to be highly visible. Seen as somewhat unusual at first, this eventually became the standard colour for all construction equipment. The diesel engine came in 1935 and model designations began RD – Rudolf Diesel's initials (others claim that the R stood for Roosevelt, the D for Diesel) – followed by a number that related to the crawler's size and engine power so that there were RD8, RD7 and RD6 machines soon followed by the RD4 of 1936. By this time the U.S. forestry service was using machines such as the Cletrac Forty with an angled blade on the front: so Caterpillar built one equipped with a LaPlante-Choate Trailblazer blade. Ralph Choate had started his business building blades for attaching to the front of other people's crawlers, his first being used on road construction between Cedar Rapids and Dubuque, Iowa.

The half-track was conceived as a way of keeping vehicles mobile away from surfaced roads where more conventional lorries soon became bogged down and its history extends back as far as the early decades of the 20th century. In the United States and Europe, manufacturers sought to produce useful half-tracks mainly for agricultural work. Holt, Nash and Delahaye were three such companies but their machines tended to be slow and cumbersome. All this changed when Adolphe Kégresse introduced his system in the early twenties in France. Having had experience of half-tracks in Russia, prior to the Revolution, he returned to his homeland where, in conjunction with Citroën, he produced the Autochenile of 1921. These machines assumed a formidable reputation, the first motor-vehicle crossing of the Sahara (1922–23) and a central Africa expedition, the Croisière Noire from Algeria to the Cape (1924–25) being among their finest achievements.

The British Army became interested and tested a Crossley lorry equipped with a Kégresse system in the mid-twenties. Due to its success, a heavier-duty system was derived from the original, making it ideally suited to military applications, and countries such as Poland (Polski-Fiat) and Belgium (FN) acquired Kégresse-type half-tracks. In Britain, Roadless Traction Limited designed numerous half-track bogie conversions for vehicles as diverse as a Foden steam lorry and a Morris Commercial. It later devised tracked conversions for Ferguson tractors and forestry tractor conversions for Land Rovers.

American and British companies were leaders in bulldozer design which initially featured manually-controlled blades and later electric control. During the twenties, Robert Gilmour Le Tourneau, an American contractor and manufacturer of earthmoving equipment for Holt, Best and later Caterpillar prime movers, developed a new system of power control which began to widen the scope of the dozer. All three control systems featured winch and cable actuation until the development of hydraulically lifted and lowered blades. One of the first British machines to be so equipped was the Vickers Vigor, developed from the Vickers VR-series crawlers, hydraulics being first used in the late thirties in time for bulldozers to make a big impression during the Second World War. Bulldozers first came to the attention of military planners after they were used to remove beach defences and even occupied pill-boxes during the campaign at Guadalcanal where Aurelio Tassone received the Silver Star for removing a hostile pill-box under fire during a beach landing. Driving up the beach with the bulldozer's blade raised to shield himself from enemy fire, he dropped it before hitting the pill-box. Later, tanks would be equipped with bulldozer blades to assist in clearing obstacles. The Caterpillar D7 saw service in all theatres of operation during the Second World War and General Eisenhower considered it to be one of the vital machines that helped win the war. In recent years the word 'bulldozer' has become shortened so that in the vernacular of the day they are usually referred to as dozers.

The Second World War interrupted Studebaker's civilian vehicle production and factory capacity was turned to the war effort. Studebaker assembled almost 200,000 US6 2.5-ton 6x4 and 6x6 trucks. Half of these went to the U.S.S.R. on Lend-Lease and the GAZ plant in Gorky produced close copies of them in the post-war years. Studebaker also produced Wright Cyclone Flying Fortress engines and in excess of 15,000 Weasels, a light, fully-tracked military vehicle. This latter machine was designed by Studebaker's engineers and can be regarded as one of the pioneers of the light crawler tracked vehicles which have become popular as down-sized agricultural machines for specialist

A JCB crawler excavator: this ageing machine is one of JCB's middleweight tracked machines and like most others relies on three hydraulic assemblies to operate as a back hoe or front shovel vehicle.

applications, ranging from viticulture to use on small agricultural sites where a conventional tractor or crawler would be too large.

Caterpillar D11R Track-Type Tractor

The D11R is the biggest dozer in Caterpillar's comprehensive range and its various blades have massive capacities, the 11SU blade being rated at 27.2cubic metres (35.5cu yds) while the 11U blade rates at 34.4cu m (45.0cu yds). These have widths of 220 and 250in (5600 and 6358mm) respectively and both are 93 inches (2370mm) high. As with the smaller models, specialist tools are also available including single- and multi-shanked rippers. These are designed to penetrate and thoroughly rip up a variety of materials.

Specification

Make   Caterpillar
Year   1996
Model   D11R
Engine   Caterpillar 3508
Power   325hp
Notes   Track-type tractor; 3F, 3R transmission; maximum speed 7.2mph (11.6km/h); track gauge 114in (2896mm); length of track on ground 243in (6163mm); length of basic tractor 6163mm; maximum operating weight 216,960lb (98,413kg)

Specification

Make   Caterpillar
Year   1996
Model   Challenger 75C
Engine   629ci (10,307cc)
Power   325hp
Notes   Operating weight in excess of 16 tons; Mobil-Trac rubber crawler tracks fitted

The amount of progress which has been made in crawler technology is illustrated by the 1996 Caterpillar Challenger which is one of a range of crawler tractors from the multi-national manufacturer whose trade-mark has become almost a generic term for crawlers.

Chapter 7
TRACTOR DERIVATIVES

Articulated Dump Trucks

Articulated dump trucks (ADTs) are trucks that have an articulated joint and oscillating ring between the cab and the body. The oscillating ring allows the cab section of the truck to move independently of the cargo body so that, even when the truck is negotiating uneven terrain, all wheels stay in contact with the ground, ensuring maximum traction as well as stability and manoeuvrability. The articulated joint allows the truck to bend in the middle and thereby serve as a steering function. The first ADTs were developed in Europe in the sixties where inclement weather, combined with confined working spaces, indicated a machine capable of carting large loads but also of being operated where the more traditional rigid-frame dump truck could not. The idea is essentially a development of the tractor and trailer combination but has grown into one of the most versatile pieces of earth-moving equipment available.

One of the first articulated dump trucks was a 4x2 machine manufactured by Northfield Engineering and displayed at the 1961 International Construction Equipment Exhibition at Crystal Palace, London. It was front-wheel-drive and had a payload of 11 tons, featured power-assisted steering and had a turning circle of 18.75ft (5.7m). The uses to which ADTs can be put is almost endless and the machines have applications in agriculture, mineral extraction, landfill, construction, and handling raw materials. The major advantages of the ADT are perceived to be as follows:

Manoeuvrability: an articulated truck's configuration allows it to turn up to 45 degrees to the right and left allowing it to be operated in tight situations. The 6x6 drive capability, along with large off-road tyres and automatic transmission, allows it to continue working in situations that may well bog down other types of equipment.

Versatility: the phenomenal growth in ADT popularity in recent years is testament to its serviceability and practicality. It can be used in areas where weather conditions are extreme, thereby extending the working season.

Ease of operation: ADTs are not complicated machines and feature accessible control panels and gear systems. The cabs are designed to provide comfort for the operator, encouraging greater productivity. Features such as automatic transmissions leave the operator free to concentrate on on-site tasks rather than changing gears and the numbers of ADTs in the contract hire market

ensure their ease of operation is a bonus to those teaching and learning how they work.

Productivity: ADTs are productive under every condition and feature the highest payload to weight ratio of any type of earthmoving equipment. Their cross-country effectiveness means that specific routes need not be planned, which is an important factor on job sites of a short duration.

Cost efficiency: the initial purchase price of an ADT is lower than that of other earthmoving equipment: fuel efficiency is high, and moving them between sites is less complex than much larger machinery because most will fit on standard low-loader trucks and do not require special movement permits.

The origins of Moxy articulated dump trucks lie in the sixties when a Norwegian, Birger Hatlebakk began to make an advanced off-road vehicle based on an articulated design which he believed would make earthmoving operations easier in rough terrain. The first experimental models required development, and in 1973 Hatlebakk's company, Glamox, acquired the rights to make another truck already in production by the Norwegian company, Overaasen. Progress was fast and the D15, D16 and D16B articulated dump trucks were eventually produced.

The company was owned by Glamox but was known as Moxy A/S. Initially the machines were made for the Norwegian domestic market but, as the concept of articulated dump trucks spread, Moxy began to export. Dealerships around Europe were established and the European market grew through the seventies and eighties. In the United States, Moxy Trucks of America Incorporated was established by Leo H. Gerbus in Cincinnati, Ohio who had 30 year's experience in the construction industry. In 1986 Moxy entered into an agreement with Komatsu Limited of Japan, which allowed Moxy trucks to be sold badged as Komatsu in markets not served by Moxy. In June 1991, the company was restructured and became partially owned by both the Norwegian mining group A/S Olivin as well as Komatsu. The company then commenced development of a new range of ADTs which included the use of proprietary components such as engines from Scania and transmissions from ZF. Current Moxy trucks are constant six-wheel-drive (6x6) articulated dump trucks in the 30–40 ton capacity range.

There are now many manufacturers of ADT-type trucks.

Specific Volvo BM agricultural ADTs of the mid-eighties included the 5350 TC 5350B 6x6. Volvo BM – Bolinder-Munktell – has refined the concept of the articulated dump truck (although it terms it a hauler) and its machines feature all-wheel drive, 100 per cent differential lock on all axles and operator-friendly cabs. The automatic transmission is designed to be fast while prolonging drive-train component wear. It is geared to maximize engine power and fuel economy. The perceived advantages of Volvo BM articulated haulers are their competent off-road ability, enabling them to maintain cycle times despite operation in rough terrain or on sites lacking access roads. The articulated steering system and rotating frame joint keep the machine operable in every kind of condition and is the key to the Volvo's off-road performance, while the design of the three-point suspension allows for high-speed load haulage on poor roads.

Specification	
Make	Volvo
Year	1996
Model	A30C 6x6
Engine	Volvo TD103 KBE
Power	216kW (289hp) @ 2200rpm
Notes	Transmission 6F, 2R; maximum speed 32.5mph (52.3 km/h); tyre size 30/65-R25; wheelbase 164in (4173mm)

Terex

The Terex 2566C ADT is based around all-welded high-grade steel frames of a rectangular box section design. The frames are engineered to exceed the stresses imposed on them when used on uneven sites. The front frame accommodates the engine, transmission, hydraulics and fuel tanks and supports the cab. Steering articulation is by two widely-spaced vertical pivot pins fitted in sealed taper roller bearings. The front suspension is intended to be maintenance-free; a leading-arm sub-frame carries the front axle and pivots on the frame. Suspension is by air-filled rubber bellows which maintain ride height regardless of load.

Specification

Make	Terex
Year	1995
Model	2566C Articulated Truck
Engine	Cummins 6CTA8.3-C
Power	190kW (255hp) @ 2000rpm
Notes	Transmission 6F, 3R; maximum

speed 26mph (42km/h); tyre size 20.50-R25;
wheelbase 160.6in (4080mm); length 372in (9445mm)

minerals need to be transported from excavations. The loader is also suitable for specialist agricultural applications, such as loading felled trees. In order to make loaders as suitable for as many applications as possible, most manufacturers offer a variety of specialized components, particularly the size and type of bucket fitted, and in many case special high lift options are available. Buckets are susceptible to wear, being constantly in contact with abrasive materials and are increasingly being designed to take replacement parts to allow for wear and tear.

Hydraulic dampers are also fitted to control movement on both bump and rebound. The rear suspension consists of axles linked to the chassis via three rubber-bushed links which provide both longitudinal location and torque reaction control. Lateral location is by means of transverse links, fitted with spherical bearings. This design ensures that the inter-axle drive-line angles are maintained at an optimum in all axle positions, ensuring longevity of the components. The loads on the rear axles are balanced by the centrally-pivoting equalizer beams. The axles are fitted with limited-slip differentials which engage automatically as conditions dictate. The centre axle has two differentials – one to drive the half-shafts for the normal cross-axle drive and the other for through differential drive. All the three axles have single reduction spiral bevel gear differentials and a secondary reduction through outboard-mounted planetary gears. Air/oil-actuated disc brakes are fitted to each hub. The whole machine is powered by a C series Cummins diesel engine and a power shift ZF 6WG 200 transmission with integral torque converter. Gear shifts are by electric control of hydraulically-operated multiplate clutches. Automatic converter lock-up engages in all forward gears to eliminate slippage losses.

Wheeled Loaders

The loader, be it wheeled, tracked or backhoe, is a machine established in the various tasks that can be loosely grouped within agriculture and is vital to many more specialized tasks, especially open-cast mining and quarrying where large quantities of

Two vintage tractors on display at the Historisch Festival Rally at Panningen, Netherlands. Opposite is a Ferguson T20 while a 1962 Hanomag Nato can be seen left.

INDEX